One of the most insidious practices in education today is the way students are denied access to high-quality mathematics instruction through the practice of tracking. This groundbreaking book not only provides concrete examples of how to make college preparatory mathematics available to all students but also shares strategies to ensure that students are successful once enrolled.

Kyndall Brown
Executive Director, California Mathematics Project
Los Angeles, CA

*A Guide to Detracking Math Courses* recognizes the urgent need to rehumanize the math classroom and make mathematics equitable for all, serving as a compelling and comprehensive guide to help educators and administrators achieve these goals. This timely must-read inspires you to develop, implement, and maintain a system that creates opportunities for learning for all.

Nancy Nagatani
Mathematics Staff Development & Curriculum Specialist
Hanford, CA

This is a book we've been waiting for! Beginning with the premise that all students are mathematically brilliant, and all teachers have strengths in teaching and learning, the book captures the essence of one team's journey to attain a detracked mathematics program. The authors provide questions to consider, reflective activities, lessons learned, and so much more to assist educators in creating equitable mathematics experiences for their students.

Shelly M. Jones
Professor, Mathematics Education, Central Connecticut State University
New Britain, CT

Detracking a school district's mathematics offerings means successfully working collaboratively through issues of policy, curriculum, professional development, and more. This book highlights the issues involved, unpacks their complexities, points to resources, and helps readers adapt the ideas to their own district context. It will be an invaluable resource for schools and districts that want to detrack their mathematics courses, in the service of equitable and ambitious instruction.

Alan H. Schoenfeld
Distinguished Professor, University of California, Berkeley
Berkeley, CA

*A Guide to Detracking Math Courses* is a well-executed guide for educators who genuinely want mathematics education reform. The detracking strategies in this book will challenge your sensibilities by stretching your comfort level with collaboration, deepening your relationships with all stakeholders, and changing the trajectory of the lives of students in a tangible way for the betterment of our entire society.

Christina Lincoln-Moore
Principal, LAUSD
Inglewood, CA

*A Guide to Detracking Math Courses* reveals the power of a shared commitment to the success of each and every student in detracked schools. More important, it offers a pragmatic and optimistic roadmap to challenging the narrative that students benefit from being sorted into courses based on past mathematics achievement. Our traditionally underserved students need advocates and changemakers; they are deserving of meaningful problem-solving opportunities that build conceptual understanding and prepare them to excel in advanced mathematics. Readers have multiple opportunities to reflect on their own presumptions about who can succeed in mathematics and what it means to be successful. It is a must-read for teachers, administrators, and community members who want and need to see the affordances of detracking.

Terrie M. Galanti
Assistant Professor, Secondary Mathematics and
STEM Integration/Computational
Thinking, University of North Florida
Jacksonville, FL

Detracking is desegregation. This is the best thing we can do for *all* math students. There should be no gatekeepers and no barriers to students learning as much mathematics as they want to.

Rori Abernethy
SFUSD Math Teacher, CTA Instructional Leadership Corps
Oakland, CA

# A GUIDE TO DETRACKING MATH COURSES

# A GUIDE TO DETRACKING MATH COURSES

## The Journey to Realize Equity and Access in K–12 Mathematics Education

ANGELA TORRES
HO NGUYEN
ELIZABETH HULL BARNES
LAURA WENTWORTH

Foreword by Robert Q. Berry III

*For information:*

Corwin
A SAGE Company
2455 Teller Road
Thousand Oaks, California 91320
(800) 233–9936
www.corwin.com

SAGE Publications Ltd.
1 Oliver's Yard
55 City Road
London, EC1Y 1SP
United Kingdom

SAGE Publications India Pvt. Ltd.
Unit No 323-333, Third Floor, F-Block
International Trade Tower
Nehru Place
New Delhi - 110 019
India

SAGE Publications Asia-
Pacific Pte. Ltd.
18 Cross Street #10–10/11/12
China Square Central
Singapore 048423

President: Mike Soules
Vice President and Editorial Director:
  Monica Eckman
Associate Director and Publisher,
  STEM: Erin Null
Senior Editorial Assistant:
  Nyle De Leon
Production Editor: Tori Mirsadjadi
Copy Editor: Deanna Noga
Typesetter: Integra
Proofreader: Lawrence W. Baker
Indexer: Integra
Cover Designer: Rose Storey
Marketing Manager:
  Margaret O'Connor

Printed in the United States of America.

*Library of Congress Cataloging-in-Publication Data*
Names: Torres, Angela, (Mathematics educator), author.
Title: A guide to detracking math courses : the journey to realize equity and access in K-12 mathematics education / Angela Torres, Ho Nguyen, Elizabeth Hull Barnes, Laura Wentworth.
Description: Thousand Oaks, California : Corwin Press, Inc., [2023] | Includes bibliographical references and index.
Identifiers: LCCN 2022059681 | ISBN 9781071880746 (paperback ; acid-free paper) | ISBN 9781071913789 (epub) | ISBN 9781071913796 (epub) | ISBN 9781071913802 (pdf)
Subjects: LCSH: Mathematics--Study and teaching-- United States. | Educational acceleration. | Track system (Education) | Ability grouping in education.
Classification: LCC QA13 .T67 2023 | DDC 510.71/073-- dc23/eng20230324
LC record available at https://lccn.loc.gov/2022059681

This book is printed on acid-free paper.

23 24 25 26 27 10 9 8 7 6 5 4 3 2 1

# CONTENTS

# FOREWORD

In mathematics education, there is a system built around the assumption that only exceptional learners can perform at the highest levels. This assumption contributes to a long-standing practice of schooling that segregates students of different backgrounds into separate experiences on pathways leading to different outcomes. Tracking (or streaming, as it is also known) of students involves segregating students for mathematics instruction based on their perceived ability, intellect, or past performance. The effects of tracking (or ability grouping, as it is sometimes called) correlate with the inequities. That is, who gets tracked into upper-level and lower-level courses, how people get tracked, and the outcomes of tracking reflect and correlate with inequities based on race, ethnicity, language status, and socioeconomic status found in our broader society. It is time to begin the courageous work needed to intentionally and systematically remove the perniciousness of tracking and the curricular and instructional practices that support this system as we move toward creating pathways for success in mathematics for every student.

As a former teacher of mathematics in urban and suburban schools, I have personally experienced and borne witness to the inequitable outcomes in mathematics learning. Too often, these inequitable outcomes result from tracking. Tracking goes hand in hand with labeling students as *high-achieving* and *low-achieving*, or worse *low-level* and *high-level* learners, further removing their perceived ability from their actual accomplishments. Such labeling negatively impacts students' disposition toward mathematics. Students perceived as low-achieving or low-level are typically segregated into low-track mathematics, in which mathematics teaching focuses primarily on rote skills and procedures. Mathematics teaching for students perceived as low-achieving does not stretch their higher-order thinking and gives limited attention to developing their conceptual understanding. Conversely, students segregated into high-track mathematics typically experience mathematics that cultivates their mathematics identities, conceptual understanding, and critical problem-solving and thinking skills (National Council of Teachers of Mathematics [NCTM], 2018). It is time to recognize and identify tracking as a systemic form of segregation.

> *It is time to recognize and identify tracking as a systemic form of segregation.*

Because of its negative impacts, it is time for schools and classrooms to detrack. NCTM's Catalyzing Change series (2018, 2020a, & 2020b) calls for the ending of the practice of tracking teachers and students into qualitatively different or dead-end course pathways. A significant body of research suggests that students' opportunities increase when schools move to detrack in mathematics. Many different research studies have shown that high-achieving students achieve at the same levels in tracked and detracked groups and all other student groups achieve significantly higher levels in detracked groups (Atteberry et al., 2019; Boaler, 2006; Horn, 2006). Detracking offers increased opportunities for higher achievement and changes in students' perceptions of themselves as mathematics learners. In addition, detracking challenges segregation in mathematics that leads to inequities based on race, ethnicity, language status, and socioeconomic status.

Detracking is often characterized as an attempt to group students heterogeneously to ensure that every student, regardless of race, ethnicity, language status, socioeconomic status, or academic ability, has access to high-quality instruction, curriculum, teachers, and material resources. While at its root, this definition is true, *A Guide to Detracking Math Courses: The Journey to Realize Equity and Access in K–12 Mathematics Education* helps us understand that detracking requires far more than rearranging instructional group patterns. Instead, it requires a holistic effort to shift beliefs, develop policies focused on detracking, implement those policies with fidelity, continuously research and monitor, and build community. *A Guide to Detracking Math Courses: The Journey to Realize Equity and Access in K–12 Mathematics Education* shows us how to do exactly that. It provides pathways for actions to move toward

- shifting beliefs about who is capable of doing and understanding mathematics;
- providing access to rigorous mathematics curricula supportive of students' demonstrating intellectual, cognitive, and cultural diversities; and
- building a community where students, teachers, and leaders feel safe and supported to engage in meaningful ways.

My NCTM president's message in June 2018 challenged the mathematics education community to move toward detracking (Berry, 2018). This book responds to the challenge by addressing the points made in that message:

- Identify, analyze, and evaluate policies, practices, and procedures to assess the impact of tracking in restricting student access to and success in mathematics.
- Provide every student access to a grade-appropriate, academically rigorous, and intellectually challenging curriculum.
- Provide students with targeted instructional time and other instructional supports to support their learning and success with a grade-appropriate, academically rigorous, and intellectually challenging curriculum.

- Analyze teacher assignments to develop balanced, supportive assignments to provide high-quality, engaging learning experiences.

- Analyze where research-informed equitable instructional practices are implemented and where not and facilitate changes.

- Provide access to mathematics coaches/specialists for ongoing real-time professional development and support, which includes but is not limited to coaching, co-teaching, co-planning, and frequent interactions on teaching and learning.

- Provide teachers and mathematics coaches/specialists with time and space to collaborate on instructional issues and continue their professional learning of both mathematics and mathematics-specific pedagogy. Teachers need opportunities to share strategies, learn new teaching techniques, meet as a department or grade level, and collaborate for improved student learning. (NCTM 2018)

Detracking is a deep commitment and investment in people, curricula, and time to support and engage every student in learning mathematics and increasing their opportunities. Therefore, teachers and leaders must be committed to the actions above when working toward the discontinuation of tracking. This book will help you realize that commitment.

Robert Q. Berry III, PhD<br>
Dean & Professor;<br>
Paul L. Lindsey & Kathy J. Alexander Chair<br>
College of Education<br>
University of Arizona

## REFERENCES

Atteberry, A., LaCour, S. E., Burris, C., Welner, K., & Murphy, J. (2019). Opening the gates: Detracking and the International Baccalaureate. *Teachers College Record, 121*(9), 1–63.

Berry, R. Q., III. (2018). *Initiating critical conversations on the discontinuation of tracking.* https://www.nctm.org/News-and-Calendar/Messages-from-the-President/Archive/Robert-Q_-Berry-III/Initiating-Critical-Conversations-on-the-Discontinuation-of-Tracking/

Boaler, J. (2006). How a detracked mathematics approach promoted respect, responsibility, and high achievement. *Theory Into Practice, 45*(1), 40–46.

Horn, I. S. (2006). Lessons learned from detracked mathematics departments. *Theory Into Practice, 45*(1), 72–81.

National Council of Teachers of Mathematics. (2018). *Catalyzing change in high school mathematics: Initiating critical conversations.*

National Council of Teachers of Mathematics. (2020a). *Catalyzing change in early childhood and elementary mathematics: Initiating critical conversations.*

National Council of Teachers of Mathematics. (2020b). *Catalyzing change in middle school mathematics: Initiating critical conversations.*

# ACKNOWLEDGMENTS

This book is a compilation of years of collaboration and collective work across one district, which includes the learnings from many knowledgeable partners in this work. Here, we thank the many people and groups who have made the work possible in San Francisco Unified and in making this book a reality. Our gratitude spans from students, to families and community organizations, teachers, site and district administrators, collaborative partners across and outside of the district, and all who have been members of the central math team. We finish with thanking our publishers, editor, and all who have helped us take our learnings throughout this journey and put them into writing. The order of these acknowledgments is not a reflection of importance since this journey was not possible without the following partners.

First, we thank the many students across San Francisco Unified School District (SFUSD) PK–12. To support all students to succeed in mathematics, we continually need to learn from our students about their math strengths, their curiosities, how they feel as math learners, and their continued experiences in math classrooms. We have learned from our students in classrooms, from their math work, on video, from empathy interviews, and in survey responses. Students have been remarkable in going about their work and allowing visitors to be intrigued by their mathematical thinking while being videotaped, during instructional rounds with leaders and visitors, while engaging in lesson study, lab classrooms, and peer reciprocal observations. We are truly thankful for the learning we have experienced from our students about mathematics, about learning, and about instruction.

Next, we thank the families and community-based organizations we worked with. We are grateful for their lens of commitment to each and every student in our community. We are thankful for parents and caregivers who are curious about how to best support their child in mathematics. We are thankful to the communities who have a particular focus on supporting all students, with an attention to equity, and have advocated for the passing of and continued to publicly support the SFUSD detracking policy, otherwise known as their Math Course Sequence Policy. These voices included members of the local NAACP, the African American Parent Advisory Council (AAPAC), Chinese for Affirmative Action, and Coleman Advocates for Children and Youth. We are also grateful for the many groups who invited us, early on and throughout this work, to share about the math course pathways and shifts in teaching and learning math more broadly to families and site leadership groups that involve families. Those inviting us in to do presentations and family math nights included local and district PTA groups, site and district family liaisons, and community advocacy groups such as our Special Education Community Advisory Committee.

The work in San Francisco was also not possible without the leadership and the input of the teachers. We are extremely thankful to all the teachers who have been willing to share their practice, ideas, activities, successes, and challenges with colleagues across the district and with the central math team so that collectively we are creating more equitable experiences for each of our students. We are thankful for the advocacy from teachers and teacher leaders at their site and through district structures and opportunities. We are particularly thankful to the teachers who advocated for the detracking policy to be passed through sending letters to the Board of Education and particularly to those who came out to the Board meeting in February 2014 to share their stories of success with heterogeneous classrooms. This work also would not have been possible without our PK–12 educators who were the foundation of the SFUSD Math Core Curriculum development, piloting, and implementation. Teacher voices have been and continue to be instrumental in supporting heterogeneous classes across PK–12.

Additionally, we have extreme gratitude for the administrators who have supported the equity work in mathematics at their sites and district level. This work was not possible without those site administrators who can discuss with parents and caregivers the reasons behind the course pathway options or without the secondary site administrators who support the collaboration of their math departments in their school schedules with structures, such as a common prep period. We are also thankful for the many district administrators: superintendents, deputy superintendents, assistant superintendents, head academic officers, executive directors, and directors, who have advocated and supported this work over time. We have been lucky that each has held an equity mindset and willingness to lead with courage and a strong voice so as to advocate for the continued support of the math detracking policy, even when the political climate has included pushback. Additionally, we are thankful for the central administrators of other central departments who have collaborated with our central mathematics team, including those from the departments of Multilingual Pathways, Special Education, Humanities, Science, Computer Science, Assessment, Counseling, Technology, teacher induction programs, and professional learning. Our collaboration across departments has created even more support for teachers and a more coherent schooling experience for our students.

SFUSD has had the enormous privilege of working alongside and learning from many collaborative partnerships over the years, many connecting to universities and other research partnerships. We are extremely thankful to the researchers and math education experts we have partnered with from Strategic Educational Research Partnership (SERP), Lawrence Hall of Science at UC Berkeley, UC Berkeley, San Francisco State University, Stanford University Graduate School of Education, University of Washington, Northwestern University, Vanderbilt University, the Dana Center at the University of Texas in Austin, UC Riverside, UCLA, Mills College, and the University of Chicago. Many of these research relationships come to us through the Stanford-SFUSD Partnership, which pairs researchers and practitioners in shared questions that are both practical and local for us as a district and also adds to the broader field of study.

Collaborations with other district teams have also been essential to our learning and sense making, leading to ideas to implement along the journey. We are especially thankful and acknowledge the learning we did collaborating with the team in the Oakland Unified School District, Lawrence Hall of Science, and SERP on our position paper and course pathways. We additionally built off of Oakland Unified's core curriculum, using their unit design and Math Teaching Toolkit to build out the SFUSD Math Core Curriculum. Additionally, participating in the many communities of practice has connected us to district leaders across California, including the Math in Common districts brought together by the S.D. Bechtel, Jr. Foundation and California Ed Partners, the California Partnership for Math and Science Education, and the California County Superintendents—especially the other member districts and counties of Region Four.

We also benefit and thrive from being part of a broader math education community, from their resources, research, and conferences. We are grateful for organizations such as the National Council of Teachers of Mathematics (NCSM), the California Mathematics Council, our local affiliate of the San Francisco Math Teachers Association, the Benjamin Banneker Association Inc., and TODOS: Mathematics for All; each has curated resources that we share with teachers and provide conferences where we have learned from and with others in these communities. Another organization that we are thankful for our continued partnership is the Silicon Valley Mathematics Initiative, which has been leading the work on assessment and peer-to-peer discourse in math for decades. We are especially excited and proud that our local and national leadership across the mathematics education community have released positions papers and policy briefs arguing strongly for expansive math in heterogeneous classrooms and against the harmful effects of tracking. None of us are alone in this work, and ending tracking is a movement.

The work on changing students' experiences in mathematics toward centering student voice in the classroom began prior to SFUSD's detracking policy. While many schools and teams and educators have been making the change in their own departments and classrooms, we want to acknowledge the larger collective teaching communities and learning that had been happening in SFUSD leading up to the larger detracking policy and implementation of the Common Core State Standards in Mathematics, providing a critical mass of teachers to whom the new policy and implementation could be spread district wide using their work as the foundation. Three specific projects happening included (1) a National Science Foundation grant called PRIME in partnership with San Francisco State University, which provided coaching and professional development to deepen content understanding for fourth and 5th-grade teachers; (2) a Complex Instruction program, in partnership with educators who taught math at Railside High School, provided professional development and coaching to a growing number of high school math departments working to re-culture math departments through focusing on the use of groupwork and the attention to status in heterogeneous classrooms; and (3) a group of K–12 school teams participating in ongoing inquiry-based professional development on Studying the Standards with the Dana Center at University of Texas in Austin and the San Francisco School Alliance. Each of these projects has connected SFUSD to expert educators in the field, and the work they began with teachers and the lessons learned through these projects created a foundation for the implementation of a detracked policy.

We would like to thank our own families and loved ones for allowing us the space to work on this book on the weekends and before and after our normal workday. Our families give us strength and support to pursue the challenging work to change school systems. While this work is not easy, it makes it that much more satisfying to be loved and cared for at home with our families.

Finally, we would like to thank everyone who helped make this book possible: Corwin for the opportunity to put our work and learnings into this book; our editors for their support and feedback throughout this process; and Erin Null, in particular, who approached us about writing this book and provided incredible guidance and encouragement throughout the writing process. A big thank-you to Robert Q. Berry III for writing the Foreword to this book, for his work on NCTM's *Catalyzing Change* that directly calls for the ending of tracking students, a resource that has been and continues to be an incredible resource for those considering, beginning, and continuing the detracking journey in mathematics. Thank you to all the teachers, educators, and colleagues who we have interviewed, to those who allowed us to reference their work, and those who read drafts and helped us clarify our ideas.

To every member of the SFUSD math team over the years, we wrote this book trying to do our best to put forward our collective ideas, questions, learnings, and activities created for educator development along this journey. None of this was possible without your voices, passion, and expertise in the work of mathematics education. We wrote this book in hopes of documenting the dynamic work needed to improve structures and systems for teaching mathematics in the SFUSD and beyond. While our journey here in San Francisco is still evolving and our work is far from over, we lift you and know that others are coming to continue this journey and build on our collective and not-yet-realized vision: Each and every being alive is mathematically brilliant, and we together can and will challenge anything that celebrates anything less.

## PUBLISHER'S ACKNOWLEDGMENTS

Corwin gratefully acknowledges the contributions of the following reviewers:

Rachel Lotan
Professor (teaching) Emerita, Stanford University
Palo Alto, CA

Terrie McLaughlin Galanti
Assistant Professor, Secondary Mathematics and STEM Integration/
Computational Thinking
Jacksonville, FL

Kimberly Morrow-Leong
Adjunct Instructor, George Mason University
Senior Content Manager at the Math Learning Center
Corwin author (*Mathematize It! Going Beyond Key Words to Make Sense of Word Problem*, 3 volumes)
Fairfax, VA

John W. Staley
Baltimore County Public Schools
Randallstown, MD

Holly Tate
Instructional Mathematics Coach & PhD Student/Alexandria City Public Schools & George Mason University
Woodbridge, VA

# ABOUT THE AUTHORS

 **Angela Torres**, MEd, is a professional learning coordinator for the UC San Diego Mathematics Project. Previously, Torres served as a math content specialist in San Francisco Unified School District (SFUSD) where she supported secondary math teachers through professional development, coaching, and curriculum support for almost a decade. Torres deeply believes in the brilliance of all students and works hard to support teachers in seeing this brilliance, including our Black, Latinx, and other students who belong to groups that have been traditionally underserved in our education system. She enjoys learning in community with teachers and has learned a tremendous amount in supporting high school teachers in San Francisco, managing the Complex Instruction program in SFUSD and through currently consulting with a team of educators supporting the Complex Instruction program in other districts. She has also presented the SFUSD math team's work at the national and state mathematics conferences. As a member of the California State Mathematics state board, she has joined a team working to support the movement toward equity for all students in California. Torres has a master's degree in education from San Francisco State University, is a Nationally Board Certified Teacher, and is always looking to bring what she has learned to the work she does with teachers. Torres's publications include California Math Council's (CMC) *ComMuniCator*, "Working Towards Equity Through Core Values," and two co-authored pieces with Lizzy Hull Barnes: a case study in *Catalyzing Change in High School Mathematics*, "Work to End Tracking and Offer Four Years of Meaningful Math Instruction," and a chapter in *Success Stories for Catalyzing Change in School Mathematics*, "Being Bold: San Francisco's Detracking Story as a Path to Equity."

**Ho Nguyen**, MEd, is the program administrator of mathematics and computer science at SFUSD. For over two decades, Nguyen has worked to support math instruction in San Francisco, first as a high school math teacher, then as a content specialist at the high school level, and now as a central office administrator. In the SFUSD math department, Nguyen supports math teachers at the secondary level through professional development and instructional coaching in addition to policy development and support. Nguyen was instrumental in beginning the Complex Instruction program in SFUSD, joining secondary mathematics teachers into an equity-centered community. He has a master's in urban education and leadership from the University of California, Berkeley. Nguyen has presented his work in SFUSD at multiple conferences including the Council of Great City Schools, California Math Council North, the National Council of Teachers of Mathematics, and NCSM: Leadership in Mathematics Education.

**Elizabeth Hull Barnes**, MEd, is the supervisor of mathematics and computer science at SFUSD. She has taught math to preschool through adults in schools in Louisiana and around the Bay Area and has supported fellow teachers and coaches in San Francisco. She believes the Common Core has provided educators a window to reframe the question, "What does it mean to be good at math?" and through her work aims to recapture mathematics as a multidimensional discipline for her district's students. Hull Barnes has presented SFUSD's work in multiple venues including the National Academies of Science, Engineering, and Medicine, the Council of Great City Schools, the National Council of Teachers of Mathematics (NCTM), the California Math Council (CMC), and NCSM. She has published articles in *EdSource* and NCTM's *Catalyzing Change*. She has also supported Stanford workshops with Professor Hilda Borko to think through the core attributes of a research practitioner partnership, considering both relationship and research design. She has collaborated with fellow math educators in multiple spaces, including intersegmental work designed with the Dana Center at UT Austin and the Conference Board of Mathematical Sciences (CBMS). Hull Barnes received her master's degree in education, with an emphasis on Early Childhood, from Mills College in Oakland, and her administrative credential from the Leadership Support Program (LSP) at UC Berkeley's Graduate School of Education.

**Laura Wentworth**, PhD, is the director of research practice partnerships at California Education Partners. For over a decade, Wentworth has worked to unite research, policy, and practice by directing the Stanford University and SFUSD partnership. She supported the development of the Stanford-Sequoia K–12 Research Collaborative and the Oakland Unified School District and UC Berkeley Partnership. She is also a lecturer at Stanford University in two courses: Introduction to Research-Practice Partnerships (RPPs) and Advanced Partnership Research. Wentworth has served on the founding steering committee for the National Network of Education Research Practice Partnerships (NNERPP), and in 2020, received the Alumni Excellence in Education Early Career Award from Stanford University Graduate School of Education. In partnership with NNERPP members, Wentworth led the development and spread of the concept of brokering across RPPs by publishing the *RPP Brokers Handbook*. She has a master's in instruction and curriculum from the University of Colorado, a master's in the social sciences of education, and a PhD in administration and policy analysis in education from Stanford University Graduate School of Education. Wentworth has articles and chapters published in and by *Phi Delta Kappan*, *Teachers College Record*, *Educational Research*, *Stanford Social Innovation Review*, Jossey-Bass, Teachers College Press, and *Educational Policy*.

# INTRODUCTION

You have likely picked up this book because the word *detracking* in the title spoke to you for some reason. We imagine that if you are reading this book, you are a leader, community member, educator, or an individual in another role in a school community who is wondering about, considering, or already detracking the mathematics program in your school or school system. You are aware of the inequities happening in your school system, particularly in mathematics, and are looking to make experiences in mathematics more welcoming and humanizing for the young people in your care, including increased access to learning and success for all students. We are excited to share this book with you, and our hope is that it offers the big ideas that help you in your own context and unique journey to advancing equity in your school, district, or state or province's mathematics education.

## WHO THIS BOOK IS FOR

We believe that all people in school systems play a leadership role. Whether you are a teacher, a site or central-office administrator (a positional leader), a math specialist, a researcher, a school board member, a community member, an elected official and policy maker, or a parent or caregiver, we know that leaders across our educational system can lead in a variety of ways. Our goal is that whatever role you play, this book inspires and equips you to examine the development of policies and practices that support detracking math classes. You may be a math teacher in a middle school, within a PK–8 district where students move on to different high school options. Or you may be a site administrator at a private or charter high school, within a system that is separate from the feeder middle grades. You may be a district leader who knows that, with the right support, heterogeneous classes would provide better access for all students, and you are curious how to start this change in your district. This book uses a combination of our expertise, existing research, evidence from practice, and the context from the San Francisco Unified School District mathematics detracking policy and story to share lessons learned about the journey to detrack math courses. It is our hope that each of you reading can consider how you might apply the lessons in this book in your contexts and in your spheres of influence.

# WHO THIS BOOK IS FROM

The four authors of this book, Angela, Ho, Lizzy, and Laura, have all worked in San Francisco Unified School District (SFUSD) for many years and have used our experiences with the San Francisco community and the stories of students, families, teachers, coaches, and administrators to support you, the reader, to do this work in your own context. First, we think it is important to give you some background on who we are as individuals to help you understand the perspectives we each bring to this work.

Angela, Ho, and Lizzy are members of the central SFUSD math team, directly working with teachers and administrators in doing this work. After working as an elementary school teacher, Laura has worked in the capacity of connecting research to practice for over a decade, specifically connecting Stanford researchers with SFUSD central leaders in hopes of helping them work on research together that is useful in SFUSD decision making and beyond. The three from the SFUSD math team have connected with Laura through various research projects throughout the years and counted Laura as an important thought partner while working to integrate research into decisions being made in SFUSD.

Angela and Ho were math content specialists as teachers on special assignments (TOSAs) in the central office, predominantly focusing on high schools. Both taught high school mathematics, previously taught in San Francisco, were members of collaborative departments, used rich curriculum such as Interactive Mathematics Program (IMP) and College Preparatory Mathematics (CPM), have taught in detracked classrooms, and have implemented group-work. After the passing of the detracking policy, Ho became the program administrator of the newly hired middle school coaching team, ranging from four to eight coaches. Additionally, Ho and Angela both managed the Complex Instruction program that started prior to the district detracking work—which is described in this book—and grew from working with a few high school teams to sites and teams across the K–12 system. Their work with this program's vision, structures, and the learning opportunities for themselves and teachers greatly influenced how they supported a district wide detracking program in all SFUSD sites through curriculum, professional development, and coaching.

Lizzy is the math supervisor and brings the experience of early elementary education, specializing in mathematics. Before she became the math supervisor, Lizzy was a site-based coach participating in the 5th-grade curriculum development group of educators that the central math team had initiated, as well as supporting fellow site-based coaches to make sense of the newly adopted Common Core State Standards for Mathematics. As the math supervisor, Lizzy has experience connecting with families, community organizations, higher district administration, and members of the district's Board of Education. Both Lizzy and Ho have led the central math team, which has consisted of K–12 math specialists and coaches ranging at times from seven to twenty educators. Collectively, Angela, Ho, and Lizzy bring the experience of a broad view of mathematics support as well as a variety of knowledge of specific touch points with elementary, middle, and high school teachers and administrators across SFUSD leading up to the detracking policy and into the present at the time of this book.

## A NOTE ABOUT LANGUAGE

As educators who want to live in a more equitable world and who are dedicated to providing more equitable experiences for all students, we are constantly considering how to be better humans, interrogating and reflecting our own practices, and thinking about how to bring ideas to others to make an impact where possible. Thus, in writing this book, we considered the language that is used with an intentionality to being the most inclusive to the best of our knowledge to all groups of people, especially in ways that highlight the strengths offered in the diversity of each community. We also recognize that each of us will choose terms to best identify ourselves, and that the listed terms we have chosen may not be the terms you choose for yourselves or your community. Here are some of the choices we made:

- When speaking of our students and naming race or ethnicity, we are choosing to use Black versus African American, Latinx versus Latino or Hispanic, and Asian versus Asian American in the spirit of being more inclusive and to reflect both those born in and outside of the United States. Because language and terminology is always evolving, you may see other terms if the writing refers to a category in previous data collection or in a pulled quote. We also tend to refer to communities that are represented in San Francisco as large focal groups. So while we may write describing Black and Latinx students as groups underserved, your context may also include Indigenous, Pacific Islander, or Arab students.

- We use the term *multilingual learner* or *emerging multilingual students* to refer to our students who speak languages other than English. Previously, the term used has been *English Language Learner*. As language has evolved, we recognize that the previous term centers English as the main language to be learned with a deficit approach to what students do not yet know and not referring to their home language(s) as a strength. The term *multilingual* recognizes the strengths students bring in their current language(s), and the term *emerging* includes the development of the English language as well as the fostering of their primary language(s).

- In one specific chapter and throughout the book we discuss *school schedules*. We use this term in substitute to what many are used to hearing as a *master schedule*. Our choice in words is because the term *master* has history in slavery.

- Throughout this book we refer to all those who have interest and influence in our school systems as referring to specific roles. These roles include students, parents or caregivers, families, community-based organizations, teachers, educators, special educators, coaches, site administrators, site staff, central administrators, Board of Education members, and so on. Previously, you may have heard people referring to *stakeholders* as a way of considering all the roles in this system. We refrain from using the term *stakeholder* given its connection to colonizing practices and the harm these practices and the term have caused our Indigenous communities (as seen in the November 2021 California State Board of Education Recommended Revisions to the 2021–2022 Local Control and Accountability Plan and Annual Update Template and Instructions).

- Throughout the book, we use the pronoun *we* to mean we the authors and we talk about the SFUSD math team in the third person to describe those who have engaged in that particular detracking effort. While all four of us were or still are core members of that team, it wasn't limited to us, and we want to honor the work of the wider team as well.

## HOW THIS BOOK IS ORGANIZED

This book begins by introducing the concept and rationale for detracking mathematics courses and walks you through ways to think about building, designing, and gaining support for a detracking policy (Chapters 1–4). Chapters 5–7 follow, sharing how to support teachers to create equitable heterogeneous classrooms through curriculum, professional development, and coaching support. Next, in Chapters 8 and 9, you will see how the structure of school schedules can support or hinder a detracking policy and how the use of research can support your work in multiple ways. Finally, the book ends with Chapter 10, which discusses how administrators and leaders can monitor and maintain a detracked policy in the future. You may choose to read this book in the order written, or you may also choose to go straight to a particular chapter. The book is not necessarily a step-by-step guide, because each of your steps will be different depending on your context. For instance, if your district or school is coming up on a curriculum adoption or considering how to strengthen your professional development model, you may begin with those chapters. If you have decided to start detracking a particular grade level or two next year, then beginning with the chapter in crafting school schedules might be a good place to start. This book gives important ideas to consider from multiple directions of how you can begin, build, and sustain this work. Each chapter starts with a story and questions to consider while reading the chapter, followed by big ideas to consider in your own journey and context, and sharing our own lessons learned from our experiences with this journey. The chapters each end with some practical questions for you and your team to think about as you move forward on your journey as well as an activity that you can implement in your work.

# THE FOUNDATIONS OF THIS WORK

We hope you notice some themes across this book that sit at the core of the beliefs that the SFUSD team has used to drive their work:

- *All students are mathematically brilliant.* This is one of the premises from the SFUSD team that you will read about. It provides the foundation to the vision of the math classroom that they are working to create across their system. It also provides the foundation to a detracking policy, knowing that heterogeneous classes can be rich environments for all students when they give opportunities for the strengths of each student to foster and shine, and where all students are able to see that they each can learn from one another.

- *Bring a strengths-based perspective to all*, particularly students, families and communities, schools, and teachers. By focusing on strengths rather than deficits, you have the benefit of improving based on the collective strengths around your community, all in service of students. Society tells us to naturally focus on deficits, so being intentional to noticing and naming strengths is greatly needed in this work.

- *Learning in community is essential to creating change.* Creating a successfully detracked mathematics system is a complex task. No one person or group has the exact process or map to make this work. We need to build off the knowledge of others within our community. This means that support for teachers and sites should be with a lens of creating strong site learning communities. This also means that from central office working to build a policy, it is essential to learn with and across departments, as well as learn with and from families and the community. Additionally. the mathematics education community spans the country and the world. Connecting and learning with and from neighboring districts and educators in county offices of education and connecting to local, state, and national mathematics education organizations and additional educational partners can all help you find new research, problem solve across partnerships, and have collaborative partners to help you see your inquiries from a different perspective.

While the San Francisco team has their own context and their own successes and setbacks, they continue to wonder how to continually improve math success for all students, continue to come across new challenges like shifting leadership and new obstacles that our students face, as well as facing push-back within the community. The hope is that sharing the story, big ideas, and lessons learned in this book continues to open the conversation across our context so that we can all continue this work and find more solutions to providing more equitable experiences for all our students.

. . . . . . . . . . . . . . . . . . . . . . . . . . . . . .

# CHAPTER 1

# YOUR GUIDEBOOK TO DETRACKING MATH COURSES

**M**s. N, a high school mathematics teacher with 12 years' experience, was excited as she walked into her new high school, ready to learn the ropes of a new school and get to know her new students. Ms. N, whose favorite subject to teach has always been mathematics, is also a passionate educator who believes all students have the ability to think and grow as mathematicians. Walking through the halls of the school, she was excited to see and hear lively and playful conversations among a racially diverse array of students speaking in a multitude of languages. Ms. N was excited to bring out her students' identities and abilities through mathematics, which was a central focus of her practice in every school she had taught in. She was also excited to join a math department that was collaborative and had spoken of a passion for creating classes that were rich in student discussions and collaboration, and where the team worked hard to create mathematics experiences that supported all students.

This year, she was teaching two different courses, Integrated Math 1 and Integrated Math 1 Honors, which happened to be the only two options for 9th graders at this school. Some of the schools she had worked at previously had as many as four different tracks for incoming 9th graders, ranging from a remedial-level course to a highly accelerated course, each of which used different curriculum and had different expectations of students, which Ms. N had found deeply inequitable. She had hoped that this new school would provide better access to deeper- and higher-level mathematics learning for all students.

Within the first few weeks of teaching both classes, Ms. N started to notice some interesting traits and similarities about her students. In her Integrated Math 1 Honors class, the students had very organized work, seemed confident in their mathematical abilities, were mostly procedurally fluent, were accurate in their calculations, and were able to complete their work quickly. It was clear to her that these students came from backgrounds in which these qualities were what was valued in someone who was deemed good at math.

They had been consistently told they were smart and had years of feeling success based on feedback and grades. But she also noticed that they seemed to have a fairly fixed mindset and became easily frustrated and embarrassed when making mistakes. They were fearful of asking questions and would only ask questions of her directly, not of each other. They had trouble articulating justifications or explanations for their work as well as when they didn't know something, which often left them stuck when attempting to solve more complex problems that required them to think abstractly or come up with creative problem-solving pathways.

By contrast, the students in the Integrated Math 1 class often had unconventional and brilliant ways of thinking outside the box when they were unsure. They asked good questions and could see ideas that others often did not see. Despite these strengths, they were also often fearful of risk taking and felt that in comparison to the students in the honors class, they were not the smart kids. Without the procedural fluency and confidence that was so clearly valued in the honors kids, they felt that math was not their best subject and had a negative disposition toward it. Often, when students would walk in, Ms. N would say, "Hi, welcome!" and a common response would be, "I hate math." Ms. N would have to follow this statement with, "Well I hope we will change that. Come in so we can have fun and do some math!"

Ms. N observed one other striking difference. In the Honors class, the students were almost all either white or Asian. In the "regular" Math 1 class, they were mostly Black and Latinx students. She couldn't help comparing the two classes and thinking that the students in both classes had strengths that students in the other class would really benefit learning from. Being new to this school and this district, she didn't know how students were sorted between the two tracks in 9th grade, but believing that all students were capable of learning high levels of math with good instruction, she knew this wasn't right. This same issue had been a challenge for her in other schools she had worked at, and she had hoped this school would be different, since the leadership seemed to take a stance in support of equitable instruction. She knew that based on the way the tracks were set up, the students who were then in 9th grade would likely stay in the same track until they graduated. This meant that the Honors students would basically be accelerated through Math 1 and Math 2 standards in their 9th-grade year. In their second year of Math 2 Honors, they would get a version of the curriculum that included all the Math 3 standards. In their third year, they would be able to take Precalculus, which meant in their senior year, they could take AP Calculus. She knew that this was a highly valued goal for some students and families in this school, given the perception that it would make students more attractive in the college admissions process.

This also meant that the students in the regular class would matriculate through Math 1, Math 2, and Math 3 in their first 3 years. Those who wanted to could go on to take Precalculus or Statistics their senior year, but *the high school graduation requirement did not require math in their senior year. The result of this track is that even for those who desired, they would not be able to reach for AP Calculus even when they were capable and interested. She could see this as plain as day when looking at which students were in AP Calculus—mostly the white and Asian students. They had access to this high level of mathematics because they had been tracked for it since 9th grade and probably earlier. She came to learn that her Math 1 Honors students were placed in that class because they had been in the honors track in middle school. And they were placed in the honors track in middle school because of how they performed in 5th grade. How was it fair that how a student learns and performs in mathematics in 5th grade should either open or limit their opportunities for the rest of their lives?*

*But what could she do about it? How could she advocate for a system within her school and beyond that would offer equal access to all students to the highest level of mathematics—a system that valued all learners as humans capable of deep thinking, learning, and growth; a system that supported all learners in feeling smart, capable, successful, and inspired; a system that valued the ideas, cultures, and languages that students brought into the building with them that could shine through their mathematics learning; a system that fostered the skills they would need in their adult lives such as collaboration, problem solving, questioning, deep thinking, and perseverance?*

The experiences and observations of Ms. N reflect challenges experienced by math educators across the United States in K–12 settings and beyond. The story also represents students' differential access to math classes and how their experiences vary based on these systems in schools. Students' experiences not only influence their access to math content knowledge and impact their *achievement* in math while in school, but they also set them up to either succeed or to struggle in their adult lives.

For this teacher dreaming of a system that served all students better—both the ones deemed exceptional and especially those who had been historically marginalized—this was an inflection point. Although Ms. N worked in a school full of teachers and leaders with good intentions, it was also a school that was not only *not benefiting* all its students, it was in fact *failing* them all in some way. All students were experiencing *tracks* that had advantages and disadvantages. Some students—though tracked to get into the highest levels of math by 12th grade—were racing through content at a fast rate but a superficial level, robbing them of the opportunity to learn to ask questions, grow from mistakes, collaborate, and think deeply about complex problems. Other students lacked access to equally rigorous grade-level coursework, lacked confidence and self-efficacy, and believed math was not a subject for them, keeping them in a track that would never give them access to higher-level math, regardless of their aspirations.

At this inflection point, Ms. N felt inspired to start a journey with her students and the teachers and leaders at her school to ultimately detrack their system. Ms. N was a dreamer, yet also a pragmatist. She knew from experience that her school didn't exist in a vacuum and that other systemic constraints would be challenges she would have to face. Not only were students tracked, really, beginning in elementary school, but these tracks also show up across the district and state systems in the form of social tracks, racialized tracks, cultural tracks, socioeconomic tracks, and privileged tracks. She knew that simply changing pathways and options of different math courses alone would not be a magic solution. She knew that realizing the dream of a system that offered high levels of mathematics instruction to ALL students would require research, advocacy, coalition building, community input, policy changes, teacher support, family support, student support, organization, patience, and resilience. But she had faith it could be done. Ms. N's experience and frustration is a common one. Her story represents that of many educators striving for equity in the math courses.

*The journey to detrack math classes involves confronting some long-standing beliefs and structures in education.*

The journey to detrack math classes involves confronting some long-standing beliefs and structures in education. Beliefs such as "Tracking helps students reach their potential" or "Only some students are gifted in mathematics" are pervasive throughout the minds of the educational community. These beliefs have led to entrenched and rigid structures in education that sort students into those who are perceived to be *good* at math and students who are perceived to be *not good* at math.

If you are looking to undertake a journey like this, this book is meant to help guide you, to help you learn from what others have done—where they have succeeded and where they haven't yet. To illustrate this, we use the story of San Francisco Unified School District's (SFUSD) journey. This journey has not been linear or perfect. It has been met with many curves, road bumps, and setbacks along the way, and the story isn't finished yet.

Your journey will most likely look different than that of SFUSD. Part of the complexity stems from the multiple parts of a school system—from policy development to community collaboration, from curriculum revision to professional development and coaching, from research to scheduling redesign—that need to change when working to detrack mathematics courses. Consider this book the on-ramp to a sometimes rocky road of changing policy and practices in mathematics. Our goal here is to help you.

## WHAT YOU CAN EXPECT IN THIS CHAPTER

*In this chapter you will*

- *Understand why detracking mathematics is an important topic in the larger context of the field of education*

- *Learn some key terms we use throughout the book and how these terms play out in practice settings: classrooms, schools, and districts*

- *Understand the prevalence of these structures for tracking in educational settings, their influence on math instruction, and what outcomes they produce*

- *Get a glimpse into the chapters ahead that will help you take steps to detrack classes in your own system, from design, to implementation, to sustenance*

## QUESTIONS TO CONSIDER WHILE READING THIS CHAPTER

This chapter explores a few essential questions, organized to provide an understanding of the larger context related to the practice of tracking and detracking in mathematics:

- *Defining key terms by talking about tracking and detracking*: What does tracking and detracking mean? How does tracking and detracking play out in mathematics?

- *Describing what tracking and detracking look like in action*: Where does tracking take place? How prevalent is detracking in U.S. schools? What makes detracking math classes so complex? What effect has tracking had on school effectiveness and student outcomes? What impact does tracking have on students' math outcomes?

- *What your detracking journey might look like*: How do the different parts of this book help me on my journey to detrack math classes in my community?

While this chapter does not provide a systematic review of the research on tracking, we refer, whenever possible, to research or cases from practice to help explain these terms. We also rely on our experiences from detracking math classes in SFUSD to provide an example of one community's journey.

## WHAT DO TRACKING AND DETRACKING MEAN?

Historically, a tracked system in mathematics classrooms has placed and sorted students into particular classes based on perceived abilities, grades, teacher recommendations, and so forth. According to Oakes (2005), "*Tracking* is the process whereby students are divided into categories so that they can be assigned in groups to various kinds of classes" (p. 3). Most often, students placed in a particular track are placed there by late elementary or early middle school, and once they are in a track, they usually do not move to another track. Consequently, tracking happens at all levels: elementary, middle, and high school. The students who are placed in high tracks are expected to be college bound. The low tracks are often at grade level or may be below grade level or remedial classes that may or may not allow a student to pursue postsecondary education.

> Most often, students placed in a particular track are placed there by late elementary or early middle school, and once they are in a track, they usually do not move to another track.

The rationale for tracking has been to provide students with instruction according to their perceived ability. In theory, the argument is that if students are placed in math classes according to their existing skill and knowledge levels, they will receive the instruction that builds on their natural ability. If students are placed in classes without the prerequisite skills, they may struggle, lose self-esteem, and may slow down the pace of instruction for the rest of the students. A tracked system assumes that some students have a more natural ability to learn math. According to Gamoran and colleagues (1995), "Ability grouping is the practice of dividing students for instruction according to their purported capacities for learning." Forms of ability grouping in math classes, like the practice of tracking, influence the instruction that students experience.

The term *detracking* means to change a tracked system of coursework that sorts students into different classes based on their perceived ability into one that places all students—with different strengths and challenges—into the same classes, where teachers use instructional strategies to support all students with their differing needs. According to Oakes and colleagues (1997), detracking involves "[moving] from homogeneous to heterogeneous instructional groupings" (p. 482). This contrasts with the notion that all students are capable of learning mathematics with the proper instruction. Students need the opportunity to access that content through instruction, what some researchers refer to as giving students the "opportunity to learn" (Carter & Welner, 2013).

Additional evidence suggests that detracking math classes is more complex than simply moving from homogeneous to heterogeneous classrooms of students at a systems level. What happens *inside* the classrooms in a detracked system also matters greatly. In a heterogeneous classroom, there are students with varying mindsets, skills, and knowledge related to mathematics. Teachers use differentiated instruction to adjust their pacing, scaffolding, and pedagogy based on students' interests, test results, and learning styles (Tomlinson, 2014). Some teachers may receive training in differentiated instruction during preservice training, and some school systems may provide teachers with professional development in differentiating their instruction. However, many teachers are not prepared for the heterogeneous classroom and its complexities.

For example, Cohen and Lotan (1997) argue that if, in a heterogeneous math classroom, a teacher's instruction does not include approaches that address issues of bias, status, and authority, then it can reinforce perceived abilities of students. Lotan (2006) describes how issues like stereotype threat—or students underperforming according to others' stereotypes of them—can influence students' performance in heterogeneous classrooms. Similarly, Domina and colleagues (2019) have gone so far as to outline five distinct dimensions of within-school, cross-classroom tracking systems to explain the variables involved, including (1) differentiation in curriculum taught within a class, (2) the level of heterogeneity in student skill levels within classes, (3) the rate of student enrollment in classes teaching *advanced* skills or skills beyond the stated grade-level standards-based requirements, (4) the extent to which students move between more or less advanced classes or tracks over time, and (5) the relationship between track assignments across subjects. The changes involved in detracking are complex and multidimensional as described by Lotan and Domina and colleagues.

## THE PERVASIVENESS OF TRACKING IN U.S. AND CANADIAN SCHOOLS

Across schools in the United States and the Canadian province of Ontario, the practice of placing students in tracks (or streams, in Canada) of classes based on their perceived ability in a content area continues to be pervasive, especially within secondary schools. In a report published by the Brookings Institution, Loveless (2013) describes the persistence of tracking, especially in mathematics, by citing survey data collected from high school principals during the administration of the National Assessment of Educational Progress (NAEP). When asked whether students are "assigned to classes based on ability so as to create some classes that are higher in average ability or achievement than others" (p. 18), more than 70% of the principals reported students attending tracked math classes from 1990 through 2011. Some professional organizations of mathematics educators have started to advocate for detracking. For example, the National Council of Teachers of Mathematics (2018) published *Catalyzing Change in High School Mathematics: Initiating Critical Conversations*, which recommended high schools in the United States stop the practice of tracking students and teachers into different math classes or pathways that do not lead to outcomes like high school graduation.

Why does tracking persist? The answer to this question is more complex. Hallinan (2006) argued in *Education Next* that teaching in a detracked school system is more difficult given the range of students' skills and knowledge in each classroom. She goes on to explain that detracking requires modifications in school scheduling and resource allocation as well as adjustments to curriculum and professional development for teachers, which may be burdensome to schools. Also, parents of perceived higher-ability students may prefer for their children to have access to homogeneous classes, which are seen as more rigorous, to prepare their children to be competitive during the college admissions process. Cuban (2018) explains the long history of tracking, starting in the 1920s, where school leaders divided students into career paths: college preparatory, general, and vocational. Then, in the mid-20th century, schools sorted students according to subject area. After research in the 1980s and 1990s from Jeannie Oakes, among others, states like California and Massachusetts started to mandate detracking in middle schools. Yet in most school systems across the nation, tracking still persists, with the rationale that students with certain perceived abilities needed access to coursework beyond their grade level.

In addition to the history of tracking in the United States, many Canadian school systems use streaming, a form of tracking, in their school system. Curtis et al. (1992) describe streaming as placing students in either informal or explicit groupings based on perceived ability. School leaders' sorting of students starts in elementary school in a more informal way through placement in programming similar to special education or gifted and talented programs. By high school, students have traditionally been placed in different streams—university preparation or *elite* streams, basic, or vocational or special education—based on their perceived ability. Clandfield and colleagues (2014) documented some attempts in Canada's school systems to eliminate streaming, or *destream*, in 9th or 10th grade to promote equity in the secondary school system. This change in streaming has allowed more low-income, Black, or Latinx students to graduate and access postsecondary education, but these authors and others (e.g., Campbell, 2021) describe how barriers still remain for students. For example, Parekh and colleagues (2011) describe how high schools in Toronto with a greater number of elite or university preparation streams or tracks had lower amounts of low-income students enrolled and higher amounts of students with university-educated parents enrolled. High schools with more vocational programming, for example, have one in five students receiving special education programming as compared to the university preparation or elite schools, which have one in eight and one in seven students receiving special education programming, respectively. Whether in Canadian or U.S. schools, streaming or tracking has been pervasive and created barriers for students historically underserved by public school systems—particularly low-income, Black, Latinx, multilingual students and students receiving special education services.

> *Streaming or tracking has been pervasive and created barriers for students historically underserved by public school systems.*

## THE IMPACT OF TRACKING AND DETRACKING ON STUDENT OUTCOMES

Tracked systems have limited math opportunities and outcomes for students historically underserved by schools. Most of the research on tracking policies demonstrates the negative effects on these specific subgroups of students because it denies them access to rigorous coursework (Cogan et al., 2001; Gamoran et al., 1995; Lee & Bryk, 1988). This in turn consequently reduces their likelihood of graduating (Gamoran & Mare, 1989), continuing on to postsecondary enrollment (Muller et al., 2010), and pursuing careers in pathways like STEM (Riegle-Crumb & Grodsky, 2010; Tyson et al., 2007). More generally, a number of studies point to the negative influence course-taking patterns have on students' achievement (Gamoran, 1997; Lee et al., 1997; Riegle-Crumb, 2006; Riegle-Crumb & Grodsky, 2010; Wang & Goldschmidt, 2003). Some studies point to the impacts of tracking that limit access to coursework by students like multilingual learners

(Thompson, 2017; Umansky, 2016) and students from low-income backgrounds, different racial and ethnic groups, and different genders (Long et al., 2012; Oakes et al., 1990; Palarady et al., 2015; Riegle-Crumb, 2006). The impact tracking has on achievement and access to coursework may be a barrier to students wishing to pursue whatever pathway interests them when they get to high school, college, career, or beyond, and more largely impacting the promise of public education to serve all students.

SFUSD provides one case demonstrating the effect of a tracked math system on students' math achievement. After years of students taking tracked math classes starting in 6th grade through 12th grade, SFUSD students' math achievement differed according to subgroups. For example, when looking at the SFUSD class of 2015 proficiency rates when SFUSD school systems used a tracked system for math classes, 19.1% of all 10th-grade students in Algebra 2 were proficient in mathematics as measured by the California State Test, while only 1.4% of African American students and 3.8% of Latinx students demonstrated proficiency on the same test by the end of 10th grade.

In addition to data showing the negative math outcomes for students in a tracked math system from SFUSD, there is an emerging research base rationalizing a move to detracked math classes in schools and districts. Some research suggests that increasing access to rigorous mathematics classes through detracking coursework and focusing on equitable practices within mathematics classrooms closes opportunity and achievement gaps. For example, Boaler and Staples (2008) describe a cross-case analysis involving three schools purposefully selected to examine equitable math teaching practices. They found one school, Railside, with detracked, heterogeneous math classes showing strong outcomes for students, both academic and social emotional, compared to two other schools using more tracked, homogeneous math classes. At the Railside school, certain conditions existed in their math classes like block scheduling with 90-minute classes, collaborative planning among teachers on a weekly basis, and content taught at a much quicker pace than in the other two schools. Railside used groupwork to structure their instruction that is related to an instructional approach called *complex instruction* designed by Cohen and Lotan (Cohen, 1994; Cohen & Lotan, 1997).

> *Research suggests that increasing access to rigorous mathematics classes through detracking coursework and focusing on equitable practices within mathematics classrooms closes opportunity and achievement gaps.*

Other case studies of schools' efforts to detrack show similar findings both in the United States and other countries. Attebury and colleagues (2019) found that detracked coursework from 6th to 10th grade and universal access to advanced coursework in 11th and 12th grades increased access to advanced coursework (in this case International Baccalaureate [IB] classes) for students who would have been traditionally placed into lower tracked classes.

Attebury and colleagues' study also found some suggestive associations between the school's work to detrack its IB classes and the closing of the Black-white gap, Latinx-white gap, and economically disadvantaged-advantaged gap on New York's Regents exams during the period of detracking.

While fewer findings exist from studies of school systems (e.g., districts and states) working to detrack their classes, some emerging evidence points in a positive direction with a few cautions. Some research by Burris and colleagues (2006; 2008) shows school districts' efforts to detrack their math classes led to improvements in students' access to higher-level math classes and increased achievement *without* negatively impacting students perceived to have higher abilities in mathematics. Some concerns with detracking are the misunderstanding and misrepresentation that it creates a ceiling for students who might otherwise accelerate their achievement in *high ability* math classes. And students in these traditionally homogeneous classrooms are now in heterogeneous classrooms, potentially constraining their ability to accelerate their learning. As you will see in this chapter, this is not the case. Detracking can serve *all* students well, for different reasons.

One notable example is McEachin et al.'s (2019) examination of California's efforts to detrack math during middle school by placing all students in Algebra in 8th grade. While controversial, implemented prior to the Common Core State Standards enactment, and generally poorly operationalized across the state, California schools' detracking effort shows some positive outcomes for historically underserved students—allowing students to improve their math test scores and access higher-level mathematics later in high school. However, other studies have found the opposite effect, with the algebra-for-all era negatively impacting student achievement in the school district they studied (e.g., Penner et al., 2015). Even McEachon and colleagues caution that the variation in implementation of California districts' detracking effort led to differential effects, with only some districts in fact placing all 8th graders in Algebra, some districts building conditions to support the teachers and students during the change to detrack the math classes. To back up this notion, another study by Domina and colleagues (2016) found variation in implementation, with some middle schools with higher socioeconomic status "tracking up," thereby increasing access for students to the more advanced math classes in middle school, including geometry coursework in 8th grade. At the same time, other middle schools from communities with lower socioeconomic status responded by being more likely to detrack as expected and placing all students in middle school within the same math classes in 8th grade.

## THE APPETITE FOR AND CRITIQUES OF DETRACKING

There may be hope that as a pathway to more equitable instruction, tracking will soon wane in practice. There are more and more teachers and leaders working toward practices and structures supporting detracking. For example, Stanford University professor Jo Boaler conducted a survey of 300 teachers attending one of her youcubed workshops, where almost 75% of respondents said they were "leading or supporting detracking of their math classrooms," or "moving to a more detracked system" (youcubed, 2021).

However, the move to a detracked system is not easy and barriers exist. Such a move will need to reflect the complexity of the process and be done in such a way that acknowledges and addresses the concerns of those who critique detracking and remain proponents of a tracked system. The move to detracking will bring and has already brought critique of the change to math course sequencing. Here are some of the questions and comments about detracking in the field that are exploring these complexities and critiques:

- Why does detracking feel like a one-size-fits-all solution? What happens inside detracked classrooms? Do heterogeneous classrooms provide access for my students' needs?

- If we detrack math courses, will this dumb down the content and slow students' progress, especially for the students who get it fast?

- Can my students still reach Calculus and get into the prestigious universities near us?

Now, let's turn to the case of San Francisco to explore some of these complexities of detracking and the answers to these questions. In this book, we use the case of San Francisco to address the concerns described previously. Other questions and comments about detracking are still being worked out by the field through further research and continuous improvement efforts in schools and districts.

## THE COMPLEXITY OF DETRACKING MATH CLASSES

In 2012, San Francisco Unified School District's Deputy Superintendent Guadalupe Guerrero knew he had a problem—one that had been going on for years. SFUSD middle schools had a two-tracked system of classes—grade-level classes and honors classes across all content areas—that was causing deep inequity in who had access to the highest level of rich mathematics learning and consequently which students scored proficient on state tests in mathematics, not only in middle school, but also beyond. SFUSD district leaders had created an administrative practice of automatically placing students who qualified for SFUSD's gifted and talented in education (GATE) program into honors classes in middle school. To get into the GATE program, students and their families had to submit a number of pieces of evidence to the school district to qualify for the program including their test scores, a letter of recommendation from a teacher, or a letter of recommendation from a parent. The district's approach to using multiple forms of evidence to help students qualify for the program was admirably aimed at increasing opportunities and removing barriers for students. Yet it didn't *always work that way. In some cases, this meant large portions of more privileged students whose families had the time, means, and knowledge to submit the paperwork could qualify for SFUSD's GATE program as a ticket into honors classes in middle school. These students had a ticket to advanced content in subjects like math and English language arts, which would give them further access to advanced content in high school, potentially even college. Yet those students whose families did not have the know-how to access what was seen as better, more advantageous classes did not have this access, regardless of their grades or mathematical aptitude or interest.*

*Consequently, the way SFUSD's GATE program operated as both a gateway and a barrier in a two-tracked system led to inequitable outcomes and failed to serve all students in the district's charge. Notably, the students left out of the GATE program, and consequently left out of honors classes in middle school, which affected their class placement in high school, tended to be Black and Latinx students.*

*When Guerrero became deputy superintendent in 2012, he was always on the lookout for what policy decisions he could make to support the district's goal of access and equity. And when he took the helm, it seemed changing this two-tiered, segregated system could help the district take one more step toward equity. At the same time, he was approached by the SFUSD math team who had been developing a new math program to meet the new Common Core State Standards in Mathematics (CCSS-M). The SFUSD math administrators reported that the new standards in middle schools pulled in content from the strands of algebra, geometry, and statistics. Specifically in 8th grade, many of the CCSS-M Grade 8 standards included what was previously in a 9th-grade Algebra 1 class and in high school Geometry. In other words, the requirement for 8th-grade mathematics was now made more rigorous and also overlapped with what was already being taught in the advanced 8th-grade algebra classes. Based on their review of the standards and their consultation with researchers and other district leaders, they believed that to teach to these new standards effectively, it no longer made sense to distinguish between what was previously divided up as 8th-grade general math and 8th-grade algebra. The new standards essentially leveled the playing field. They would need to eliminate the two-tracked system of general math classes and honors math classes in all SFUSD's middle schools. Could this be the moment Guerrero and other SFUSD leaders had been looking for where they could create a more equitable system of accessing classes in SFUSD middle schools?*

> *The process of planning, developing, implementing, and supporting a detracked system cannot be rushed. It needs to be purposeful, methodical, and well-supported to succeed.*

Similarly to the complexity presented in this story about SFUSD's work to detrack its math classes, a lot of the research we have summarized thus far has emphasized the complex endeavor of detracking. For example, Burris and colleagues' study (2008) discussed the importance of conditions such as high expectations, beliefs students can achieve, and enriched curriculum to support the detracking efforts. The process of planning, developing, implementing, and supporting a detracked system cannot be rushed. It needs to be purposeful, methodical, and well-supported to succeed.

The story of SFUSD's effort to detrack its math classes is also complex. The process did not involve a simple change in school scheduling. To illustrate that complexity in SFUSD the team working to detrack attended to three key design features they believed were necessary to support a successful detracking effort:

1. *Systems and policies designed to support equity and access.* To achieve an equitable and accessible system of math instruction that supports all students, including historically underserved student groups and historically well-served groups, to be successful and learn at a deeper level of understanding, there needs to be a systems-level policy change that influences central leaders' decision making, school site leadership and operations, classroom practices, and community mindsets. The SFUSD Board of Education adopted a new policy in 2013–2014, which was developed using a couple of steps. First, the school district leaders developed a partnership with the Strategic Education Research Partnership (SERP), an intermediary nonprofit organization centered on bridging research and practice, who helped engage with experts from the field of mathematics over a 10-month period to help think through the design of a systems-level policy that could enable CCSS-M implementation. Some of the concepts that formed from these discussions were

- the need for heterogeneous classrooms
- at least one course pathway in high school that students can opt into to reach Calculus by 12th grade
- decision points along the way for students and families to choose math class pathways based on the students' interests

Leaders in the effort used information from their research partners, local experts, and other sources like SFUSD teachers' math expertise to articulate the rationale for the policy outlined in a position paper that defined the class pathways needed to support the CCSS-M. The thought partnership with these researchers and other local leaders and experts bolstered SFUSD leaders' understanding of the complexity involved in detracking math classes. In addition to confidence, the position paper provided common talking points with evidence for leaders to reference when explaining the reason for the policy change, including using SFUSD's own historical data showing incredible inequities and opportunities gaps over years of mathematics instruction.

## LESSON LEARNED

Thought partnership between local experts and SFUSD leaders bolstered leaders' understanding and confidence of the rationale for detracking math classes while also building a body of evidence to support the policy and practice changes.

2. *Math instruction within diverse heterogeneous classrooms of students*. At the heart of the journey to detrack math classes sits (1) the enhancements of curriculum, (2) professional development, and (3) coaching combined with (4) a change in policy related to the math classes sequence. These four levers, when interconnected and designed with equity in mind, support the development of effective teaching and the subsequent learning for all students, especially those historically underserved.

To support a detracked math classroom, teachers will need to teach heterogeneous groups of students, whose backgrounds and experiences, as well as mathematical strengths and understandings, differ. What is the vision for a heterogeneous math classroom? For San Francisco, teachers created a vision statement of the equity-based math instruction that they are working toward in all classrooms: *All students will make sense of rigorous mathematics in ways that are creative, interactive, and relevant in heterogeneous classrooms.* This statement and all the four levers supported educators, administrators, caregivers, and community members to consider what it might look like, sound like, and feel like to be a math learner in any San Francisco PK–12 math classroom. The central math team created guiding principles (see Chapter 3) and premises (see Chapter 6) to help others expand their understanding of this vision, expand on the many ways people can all be smart in mathematics, and view each student with a strengths-based lens of knowing that each brings brilliance to the mathematics.

Before beginning its detracking journey, SFUSD had strengthened its equitable math instruction practices through various professional development programs, even when resources were scarce. One of the professional development programs in SFUSD was specifically created to re-culture mathematics departments through the lens of Complex Instruction (Jilk & O'Connell,

2014). Based on research by Elizabeth Cohen and Rachel Lotan (1997) at Stanford University, Complex Instruction develops teachers' instructional skill in teaching heterogeneous classrooms by using strategies that mitigate status issues, often stemming from racial and social power imbalances in classrooms, to improve access to participation with the mathematics content (Cohen, 1994). This equity-centered approach became a foundational step to creating a critical mass of teachers who had successes with heterogeneous classes and were an important part of the conditions that allowed SFUSD to successfully detrack its math classes.

## LESSON LEARNED

Detracked math classes require a strong network of learning for teachers and leaders to support rich instruction for all students and enact the curriculum needed for heterogeneous math classes.

3. *Leadership at all levels of the system advocating for the same change.* In addition to these important conditions—curriculum, professional development, coaching, and ultimately the operations involved with changing the math sequence—as well as the systems-level policy change focused on equity, San Francisco's change to equitable and accessible mathematics also took specific leadership moves. To pull off a wide scale change, SFUSD would need leadership from students, teachers, coaches, site leaders, central leaders, and policy makers articulating and defending a new approach to mathematics instruction, even in the face of the resistance from some communities to make this change.

Leadership looked like everyone from superintendents Richard Carranza and Vincent Matthews to SFUSD math teachers and students advocated for the same policy change in concert at the right moment. The support for the policy started with Superintendent Carranza but continued across changes in senior leadership from the chief academic officer to the superintendent. Carranza gave an impassioned speech to the SFUSD school board and subsequent SFUSD community. As seen here, Carranza quotes specific statistics about which students take Advanced Placement math exams (either AP calculus or statistics):

> *Of the 928 students who took those AP math classes at this high school over the past two years, <u>only 7 were African American, and only 21 were Latino</u>—NOT 7% and 21%, but **<u>7 and 21 actual students.</u>** (At this school there was an approximately 10% Latino student population and a 3% African American student population.)*

In other board meetings, many teachers advocated for the new math policy and the increase in heterogeneity in math classes. Most of these teachers had been trained in Complex Instruction, an approach to instruction that centered on heterogeneous classrooms. The teachers also became the early adopters of the SFUSD homegrown curriculum and took on leadership roles in the teacher professional development.

## LESSON LEARNED

Many leaders at all levels must be on board with the change and be able to sustain over time their defense of the change to detrack the math classes to make the most persuasive case to all community members.

While this was SFUSD's journey to address the complexity of detracking math classes, your school, district, or state may have different conditions or realities that make your journey a bit different. As you continue to read through this book, we come back to the SFUSD's story as an example of a school district working to detrack its math classes. We share lessons learned along the way, and hopefully SFUSD's story can provide some inspiration and ideas to support your journey.

## USING THIS BOOK TO SUPPORT YOUR DETRACKING JOURNEY

In this book, we walk you through a road map to detrack math classes and help you think about—and even plan out—the necessary conditions along your journey. Think about this as a workbook that will help you put into place both ideas and specific practices for detracking within your school and district settings. This book is divided into three parts: developing a policy for detracked math classes, implementing detracked math classes, and sustaining detracked math classes.

### Developing a Policy for Detracked Math Classes

The first part will support your understanding of the policy development process to detrack your math classes. In Chapter 2, you will explore the different levels of the context in which your detracking efforts will take place. As you develop the policies to guide your detracking of math classes, you will need to take into account the realities of your unique context—the leadership, the history, the goals, any legal action, and so on. In Chapter 3, you will explore how to design a policy to detrack your math class. You will work on building the vision for a detracked system of math classes among all community members and outline your visions for the future of math classes. You will explain your *why* or the rationale for this change and explain the steps you took to reach this conclusion. You will also explain the investigations and evidence collected to inform the design of the new policy. In Chapter 4, you will talk about working with all the community members to gain support for the policy. You will work with the students, families, teachers, school leaders, district leaders, school board members, policy makers, and other administrators necessary to gain their support, or in some cases prepare for their opposition to your policy. We explore building coalitions to get the policy passed by whatever governing agency—instructional leadership team, school board, or state congress—needs to approve the policy.

## Implementing Detracked Math Classes

The second part of this book helps you design the implementation of your detracked math system. Chapter 5 explores the selection and design of curriculum to support student collaboration in heterogeneous classrooms. Chapter 6 addresses the necessary professional development for teachers to support their instruction in classrooms requiring dynamic instruction. Finally, Chapter 7 highlights the role of instructional coaching during the implementation of new heterogeneously grouped math classes as another means to support teachers' professional learning. These chapters explore the important conditions that will allow teachers and leaders working in the new detracked math classes to thrive.

## Maintaining Detracked Math Classes

In the third part of this book, you will explore how to sustain a detracked math system. You will examine the role of research in Chapter 8 in understanding the implementation, development, and impact of a system of detracked math classes. In Chapter 9, we discuss monitoring the policy by collecting internal evidence using continuous improvement practices. In Chapter 10, we discuss how to support ongoing stakeholder engagement as one approach to maintaining an effective detracking policy.

► As we compile the lessons learned from research, the experiences of the team in SFUSD, and cases from other schools and districts, we want to offer a set of questions that the team found helpful as they started their journey to detrack math classes. These questions can be used to help you prepare for your professional journey as well as in small group discussions related to planning for detracking. These questions relate back to the SFUSD journey, but we think they are generalizable in nature:

- What are the demographics of your students, teachers, leaders, policy makers, and community members? What is their racial or ethnic identity? What special programs do they participate in (e.g., programs serving multilingual learners or students with an Individualized Education Program)?

- What are various community members experiencing from their perspective as it relates to math classes in your school, district, or state? Do certain groups of families currently favor or oppose your math programming as it is currently designed?

- From your perspective, what are the conditions that could support detracked math classes? Currently, how does the leadership at all levels, the professional content and systems for learning, and the current policies and structures support detracked math classes?

# Activity 1: Starting Your Journey With the Five Whys Protocol

▶ **Directions:** Use this protocol to explore the rationale for why your community would want to detrack its math classes. You can do this activity with a large group that splits into small groups (3–7 people) to work through the questions and share out their responses, or do this with a smaller group (3–7 people) that works through these questions together.

- Step 1: **State the problem.** What is the problem your school, district, or state is trying to solve by detracking math classes? See if you can relate the problem back to student outcomes in mathematics.

- Step 2: **Ask why the problem is occurring.** Anticipate having your team or teams name multiple reasons why.

- Step 3: **For each problem identified, ask "Why?" up to four more times.** Select one of the reasons why, and then ask the team to come up with another explanation or rationale. Continue to ask why until there are no more reasonable explanations to the question. You may go through this process a few different times based on the number of original reasons your team came up with in Step 2.

- Step 4: **Agree on an action to address the problem from occurring.** In some cases, your team may leap to, "Let's detrack our math classes." This is good, but detracking is complex. If you make this leap, then ask your team members to go back to the why to see if there are a few specific actions that could address the multiple causes to the problem unearthed by the Five Whys protocol.

# PART 1

# DEVELOPING A POLICY FOR DETRACKED MATH COURSES

# CHAPTER 2

..............................

# GETTING TO KNOW
# YOUR POLICY CONTEXT

*I*n December of 2008—less than 2 years into the tenure of Superintendent *Carlos Garcia—the San Francisco Unified School District (SFUSD) school board passed an ambitious policy aimed at improving access to college-ready coursework. The policy changed SFUSD's graduation requirement by aligning it with the entrance requirements with the state university entrance requirements. This high school graduation policy was part of an ambitious agenda to support the goal of access and equity in the school district's strategic plan written by Garcia and his team titled, "Beyond the Talk." The goal reads:*

> We believe access and equity are at the heart of making social justice a reality. The politics and ideology of social justice are empty without daily actions that improve the living and learning conditions for the children of San Francisco. Do our teachers have a broad range of teaching styles and skills to draw on; are they fully aware of current research on human development; do they know their content deeply; and are they able to know all groups of students, including our target groups (African-American, English Learner, Latino, Pacific Islander, Samoan, and Special Education students) while also knowing the unique gifts and talents of the individual? For some the answer is no. Our answer has to be YES, YES, YES, YES! We must create an organization that ensures every student has access to these capable teachers, and we must be an organization that knows and supports teachers. Authentic access and equity will exist when our families, students, and teachers report that they are clear about what is expected and have the support they need to meet those expectations.

*This district goal that focused on access and equity and supporting more students to meet the high school graduation policy became a rallying cry for several key policies passed by the SFUSD school board. It paved the way for what happened when Richard Carranza succeeded Garcia as superintendent and Guadalupe Guerrero became deputy superintendent in 2012. Together with the SFUSD math team, these leaders saw detracking as a natural enactment of this equity and access goal and worked toward the passage of the detracking policy that began in 2014. Like SFUSD, each school and district has a history of key conditions that influence the context for any policy it adopts or changes. We refer to this reality as the policy context influencing the development, passage, and enactment of policies within a school or district.*

## WHAT YOU CAN EXPECT IN THIS CHAPTER

*In this chapter you will*

- *explore the context that will influence the development of a policy and practices aimed at detracking your math course(s), including key variables and conditions involved in the context of a classroom-, school-, district-, or state-level policies related to math course policies*

- *learn how to conduct a premortem (rather than a postmortem) of your policy context by discussing some essential questions and explaining key lessons learned from SFUSD*

## QUESTIONS TO CONSIDER WHILE READING THIS CHAPTER

Here we present a set of questions that will help you think of your own policy context while you read this chapter. These questions relate to some of the variables in your policy context that may influence your work to detrack your math courses.

- **Leadership:** What is the tenure of key leaders in the classroom, school, district, school board, or state? What are their held beliefs about detracking?

- **Goals:** What are the goals of the classroom, school, district, or state? Would detracking help your school or district achieve those goals?

- **Legal decrees:** Is there specific litigation or historical legal decrees requiring specific policies or practices in your school or district? What influence might these legal decrees have on efforts to detrack math courses?

- **Existing policies or practices:** Are there federal or state requirements influencing policy development or enactment? How do these requirements relate to detracking efforts?

- **Influential community groups:** What is the role of the teachers or administrators in your school or district as they relate to detracking? Is there a union for either of these groups, and do union leaders hold any beliefs about detracking? What role have family groups played in the development of other policies? How do teachers and families view detracking courses, especially families of minoritized communities?

- **Historical events:** Are there events that happened in the past within your school or district that influence the organizational memory of your classroom, school, district, or state? For example, was your school or district involved in desegregation efforts after the passing of *Brown v. Board of Education*?

As we write in more detail about all the variables within a policy context, keep in mind these questions, which we believe highlight some of the more influential variables influencing any work to detrack math courses.

## DIFFERENT VARIABLES WITHIN YOUR POLICY CONTEXT

When classrooms, schools, and districts make large-scale changes such as a detracking policy, they are not doing so in a vacuum. These changes happen within a larger context that exists across multiple levels—from the classroom, to the community, often even to the state and national level. Here we describe specific variables within a school or district context that will influence your work to change a policy or practice. While many of these variables are explicit, there are some implicit variables such as biases of people involved in the school system or unwritten power or lack of power people have because of their racial, gender, sexual, or other identities.

### Student, Staff, and Classroom Context

At the classroom level, teachers and students adopt and implement policies and practices that guide their structures and interactions in the classroom. There are policies and practices for teacher-student interactions and student-student interactions. There are policies and practices for completing assignments and administering tests. Classrooms also have sociocultural realities related to each student and each teacher or teachers: their race, ethnicity, language, gender identity, religious affiliation, immigration status, sexual orientation, and other ways of identifying and belonging.

Some social psychological factors of the students and teachers may influence a classroom. There is some research about teacher biases that influence their perceptions (e.g., Weathers, 2019) and behavior (e.g., Okonofua et al., 2016) toward students with different racial or gender identities. Similarly, students' mindsets (Dweck & Yeager, 2019), stereotype threat—people's perceptions that they may conform to a stereotype (Steele, 1997)—and sense of belonging (Walton & Cohen, 2007) can influence the student, staff, and classroom context.

The classroom is also influenced by who is chosen to teach in the class. School leaders, such as principals and assistant principals in the United States, are usually tasked with hiring and supporting teachers. Some research has associated greater growth in student achievement and well-being with different teachers and characteristics. For example, novice teachers are often assigned to classes with students potentially needing more experienced teachers (e.g., Kalogrides et al., 2012), while other research shows teacher effectiveness and experience support student learning (e.g., Podolsky et al., 2019).

## School Context

At the school level, policies exist to guide student and staff conduct, what counts for student attendance, and how to safely be dropped off and picked up from school. There are some school-level policies influencing instruction, such as the school schedule—when certain content is taught and when students attend certain *special* classes in elementary school like art or music. In middle or high school, policies related to school scheduling influence how students and teachers are assigned to certain courses.

Schools are led by principals. Each school usually has goals set by its leader and its community members, which are usually unique to the community based on the needs of the students and staff, such as building the achievement and engagement of multilingual students and families or implementation of new instructional strategies. Also, secondary school departments have teachers who are department chairs whose decision-making influences which teachers teach upper-level classes, like AP or IB classes, and the department-level goals or capacity-building efforts within the department. Other leadership positions like counselors or student advisors influence students to take certain classes.

In school contexts, there are historical events that influence policy. For example, some schools may have a history of violence, some may be among the first schools to implement desegregation programs, and some may simply have a special alumnus like a former president, athlete, or other well-known figure. Similarly, there are schools with a history of being *elite* and having special criteria for admission (written or unwritten). Or some schools may have a history of being open to social justice and reform, like being named after social justice activists like Dr. Martin Luther King Jr. or Harvey Milk.

Schools also have a sociocultural context influenced by the identities and demographics of students, family, and staff as well as the surrounding community. The sociocultural context can influence the socioeconomic status of the students and families or the languages spoken by students and staff. Schools' status as either private, public, or a publicly funded charter school may influence the sociocultural context based on their admissions requirements. Private schools also have a financial commitment as well as admissions criteria, public schools may have geographical attendance boundaries, and charter schools may have a lottery system that influences the admissions to the school community and consequently influencing who makes up the social and cultural demographics of the community.

## District Context

At the district level, policies guide requirements for enrollment in schools and high school graduation. Districts have policies and practices for school start and end times, required minutes of instruction, requirements for assessment or grading, and local standards, scope and sequence, and curriculum purchases. Standards and curriculum are especially influential over math programming in districts. Generally, standards and curriculum may lend for more or less tracking depending on how narrowly defined the mathematics is—easily separating those who are seen as smart and those who are not—or how expansive the mathematics is—allowing for there to a multitude of ways to approach the mathematics and many ways for all students to access the content and learn the depth of the content.

School districts have state policies, federal policies, and some legal decrees that dictate how they operate. Districts have rules from the state for how to allocate funds in their budget. States also give school districts certain standards to guide how they teach special populations like multilingual learners. For example, if your school or district wants to detrack, you will have to account for your state's English language development standards in the instruction of your more heterogeneous mathematics classes. Policies in school districts are also related to federal policies such as requirements from the 1995 Individuals with Disabilities in Education Act (IDEA) or the 2015 Every Students Succeeds Act (ESSA). Districts also follow policies dictated by litigation articulated through consent decrees. These legal decisions may guide how school districts appropriate funds in their budgets or how they instruct certain populations like multilingual learners.

School districts are mostly led by democratically elected school boards, but sometimes they are led by a mayor or a state-appointed representative. The leaders of the school district hire the superintendent of the school district to run the administration. They also set policies and budgets, and they hold the administrators and educators running the schools in the district accountable for achieving goals.

Also, it is important to note school districts come in all different forms. Some are like SFUSD, which is a unified school district, Pre-K through 12th grade; some are elementary districts with kindergarten through 8th grade; and some are high school districts Grades 9 through 12. Also, other school districts have other overlapping bureaucracies like early education programs that are funded by Head Start or county education programs like adult education or education of students in juvenile justice programs. For example, SFUSD is an organization that functions as a school district and a county office of education, which creates some overlapping bureaucracies, whereas with other school districts, the county office of education supports multiple school districts.

## Community, State, and Federal Context

The local community, including the city, county, and broader community, influences schools. City and county governments influence policies on school safety and whether schools can be open given public health realities (as seen during the COVID-19 pandemic). Many states have county offices of education that are meant to help districts with technical assistance and set and meet their goals related to accountability. Also, many schools are influenced by community organizations. There could be a nonprofit organization that specializes in teaching math in a certain way or organizations representing special interest groups, like the NAACP or other racial, cultural, or gender groups, who might advocate for better math instruction for their special population, or for more inclusive practices across populations.

However, the state context may play a more dominant role compared to the city and county context, especially in relation to detracking efforts in mathematics. For example, state decision makers set minimum graduation policy requirements or adoption of certain curriculum, standards, or assessments for accountability purposes. For example, when many states adopted the Common Core State Standards, for some districts like San Francisco, this change became one of a few different catalysts that led to the tipping point for making changes to their tracking system in mathematics (this is discussed more in the chapter on curriculum). Further, there can be state-level litigation that requires schools and districts to implement specific practices. For example, states dictate the requirements for teaching English language development for teaching students designated as English learners. States may also require more funding for subgroups of students or schools producing certain levels of achievement.

While the federal government has less influence than that of state-level policies, there are some federal requirements that influence all schools and districts, such as reporting requirements for receiving Title I funding through the Every Student Succeeds Act. State-level policies are influenced by these federal requirements, such as making sure to assess students in reading in relation to federal Title 1 funding.

## ANTICIPATING VARIABLES THAT COULD INFLUENCE YOUR MATH COURSE DETRACKING POLICY

We want to introduce you to the concept of a premortem—a project management strategy where you anticipate a future policy failing or not achieving its goals and imagine what variables may undermine the policy. Let's imagine you are a teacher or leader, and you implemented a new detracking policy at the classroom, school, district, or state level. The policy has not achieved its intended goals—to provide equitable access to math courses and instruction to all students through classrooms that are filled with students who have a variety of skills and knowledge in math, sometimes referred to as *heterogeneous classrooms*. So far, the policy has not produced its intended results—students more engaged in math instruction, increases in student grades and course attendance, and students having more access to math content within heterogeneous content—than previously reported, but why? Let's look at some possible reasons.

## Leadership

As described in Chapter 1, in the United States and to some extent in Canada, most classrooms, schools, districts, and states support a tracked course system in mathematics. Tracked systems of math courses reflect the belief that at a certain age, sometimes as early as elementary school, there are some students who are perceived to be simply better at math than others. Consequently, more *advanced* students are put in classrooms and given instruction that is perceived as moving through more advanced content at a faster pace than that experienced by their peers. A policy that detracks math classrooms goes against this status quo. School leaders may question when a teacher, principal, or policy maker wants to develop a policy that goes against this system of belief and supports a different math course sequence. When a leader questions the new policy, this can undermine the support of others such as students, families, teachers, other leaders, or policy makers. Conversely, when leaders support a move against the status quo, it can help give an effort the fuel it needs to win over other interested parties.

You may be a teacher working in a classroom with heterogeneous groups. You may encounter leaders who question why you are not splitting your students into separate groups by ability. You may be a secondary principal who does not separate students into high-track or low-track math classes in 9th grade, and then sorts students into a general math and advanced track in 10th through 12th grade, and district administrators may question why you are not providing *advanced* students the coursework they need in 9th grade. Your premise may be that all students are capable of achieving and you prefer not to label some students as advanced or not advanced, and students needing advanced coursework could have access to this type of content when differentiated instruction is employed by teachers. You may be a district leader who is implementing a policy that detracks math classrooms in middle school. You may encounter families who want their students to receive more specialized math instruction early rather than being in general math courses in hopes that they can reach the highest levels of mathematics in high school.

Research suggests that most reforms at the school and district levels are heavily influenced by leadership decisions and actions (Childress et al., 2007; Zavadsky, 2012). For example, Bryk and colleagues (2010) found school leaders' actions were one of five essential supports needed for a school to effectively serve all its students, with leadership acting as the catalyst for major changes at a school. Similarly, Zavadsky (2009) found that systems and conditions created by district leaders in districts seen as effective heavily influence how effectively a school system serves different groups of students.

In SFUSD, it was essential that the leaders within the school at all levels—from teachers in the classrooms to the superintendent—advocated for the detracking policy. At the time, SFUSD superintendent Carranza gave a speech to the community during a school board meeting advocating for the policy. Imagine if he had made a speech questioning the policy. If the latter were the case, the policy supporting detracked math courses through 10th grade would most likely not have passed.

Further, this work is complex and requires a team of leaders to help in the effort. In the case of SFUSD, Superintendent Carranza empowered a chief of staff to develop funding to support the implementation of the math policy, including a lucrative, long-term grant from the Salesforce Foundation. The head academic officer worked with STEM Executive Director Jim Ryan, and they were instrumental in designing and encouraging the new math detracking policy because they led the systems-level planning, communications, and implementation explaining the purpose for the change in math courses in SFUSD. Finally, the school district math leaders, three of whom are authors of this book, played the role of representing the work to other district leaders who are their peers in the curriculum and instruction department and principal, supervisor department, the school principals, and the teachers who all needed communication about the change in the math policies and practices focused on detracking.

## LESSON LEARNED

Detracking efforts are best achieved when all leaders and community members are communicating an aligned vision and leaders collaborate as a team on the many facets of the effort.

## Goals

Policies often reflect the status quo of a school's or district's goals, and as with leadership, policies seen as opposing the status quo may meet with opposition. In the United States, most school systems work to accommodate students who are seen as having advanced math skills through a myriad of course pathways, including honors and Advanced Placement classes. In fact, secondary schools are often ranked by how many advanced courses they offer to their students. Consequently, classroom, school, district, and state goals may aspire toward outcomes that privilege students who have more skills than other students. Often, classrooms with students working above their grade level receive more experienced or higher-skilled teachers and may even receive access to more opportunities like taking an Advanced Placement test in mathematics that could give students college credit for a high school course.

Teachers may be asked to meet these types of goals when their administrators require them to create specialized programs for students who are seen as gifted at mathematics rather than creating math programs with the premise that all students are mathematically brilliant. School leaders may be asked to meet these goals when district leaders ask them to meet the needs of every student, including students perceived to be more advanced in math than others. District leaders may experience this goal because they are expected to design gifted programs based on state-level mandates or laws. But often having a focus that prioritizes gifted or advanced students means deprioritizing the support for other students. Too often students not in these special programs are seen as less capable, less worthy of the most excellent teachers, and less interested or able to handle the rigor of mathematics that would offer college credit or guarantee college placement.

SFUSD took a different approach to prioritizing programming only for students deemed gifted. Instead, SFUSD was one of the first school districts in the nation to name *equity* as a goal in its strategic plan and recognize that all students deserve access to rigorous mathematics instruction. The district leaders adopted the premise that all students can rise to the level of challenge and that all students should be encouraged to have a strong mathematical identity and self-efficacy. By naming this goal, the leaders of SFUSD were saying they were going to focus their resources on students who have been historically underserved or overlooked by most school systems. One way to do this was to design programs that *center* the needs of historically underserved students, while *also serving* the needs of all students in SFUSD. As you will read in subsequent chapters, SFUSD work on its math program was also meant to help all students learn according to the new Common Core State Standards in mathematics, and the design was meant to change structures of instruction that privileged some students in mathematics over others.

> *SFUSD was one of the first school districts in the nation to name equity as a goal in its strategic plan and recognize . . . that all students can rise to the level of challenge and that all students should be encouraged to have a strong mathematical identity and self-efficacy.*

## LESSON LEARNED

Goals related to your math detracking policy need to be aligned to broad institutional goals that (hopefully) center concepts like equity and access.

## Legal Decrees

Historically, educators are often bound to follow certain policies that stem from litigation, regardless of how that affects their desired use of resources. Moreover, adopting certain education policies may incite scrutiny from others in the community and may produce fresh legal action against a school or district. There may be legal decrees and oversight practices that could restrict or enhance the design of a math detracking policy.

A great example of this in the case of SFUSD had to do with existing consent decrees. When SFUSD leaders were working to pass their new math course sequence, SFUSD was still governed by a specific consent decree that influenced the way the school district taught mathematics. The consent decree related to the *Lau v. Nichols* lawsuit. What is known in SFUSD as the *Lau Plan* was a consent decree working to improve how SFUSD schools support multilingual learners and their language development. For elementary schools, this meant they were required to have a daily 30-minute English language development period for students designated as English learners. For district leaders, this means they had certain systemwide practices they had to hold people accountable to implementing, and they had to liaise with a court-ordered monitor. For the SFUSD leaders crafting the new math detracking policy, this meant they would have to build a policy accounting for the support needed for students designated as English learners. Also, district leaders would need to provide additional coaching and professional development to secondary math teachers who were teaching in more heterogeneous classrooms with a broader mix of multilingual learners at various stages of language development.

Some researchers see litigation as one of the important levers the community has for improving K–12 education, while others question the resulting changes from legal action. For example, Koski has argued that legal action is an important influence for maintaining equity for educational funding at the local and state levels (Koski & Levin, 2000; Koski & Reich, 2007). Conversely, Hutt and Polikoff (2020) point to some inconsistencies in the results of legal action, agreements, or settlements. In arguing for a new public accountability framework for public education, they point to the *Williams v. California* settlement—which required schools and districts across the state to report student access to resources like textbooks—as a case where a legal settlement resulted in a flimsy implementation of the desired goals of the settlement. Hutt and Polikoff point out the inconsistent data in schools' reporting and the lackluster public scrutiny of these reports. While imperfect, legal action can be something schools and districts need to pay attention to and may influence their policies.

## LESSON LEARNED

Active or historical litigation in your school or district could relate to your math programming. If legal decrees exist, they could influence your detracking efforts. It is important to know your legal landscape.

## Existing Policies and Practices

There are instances where federal and state policies require educational leaders at the county, city, district, and building levels to maintain certain local policies and practices. For example, the state policy that influenced math detracking in SFUSD was California's adoption of Common Core State Standards in Mathematics. Like many districts and schools around the country, when SFUSD leaders examined the new standards, they saw a separate set of standards for 8th-grade math and high school algebra, yielding two courses where there was previously one Algebra 1 class in 8th grade. This started the conversation in their math department about the types of course sequence needed to enact the new standards.

While power to enact education programming sits in the hands of state leaders, federal leaders can also make policies related to their federal funding like Title I funding or other federal incentive grants. In 2009, the Department of Education launched the Race to the Top program, a grant that states competed for. Some states passed policies such as requiring teachers to be evaluated using standardized test scores as one of the outcomes examined in the accountability system (e.g., Baker et al., 2013).

For teachers, you may experience these state and federal policies from the time you are getting your teacher certification to how you are evaluated by your principal. School leaders, you may feel these state and federal policies every time you set and report progress on annual student outcomes or meet with your school site council to review those goals. For district leaders, you might experience these state and federal policies regarding how you are expected to spend federal funding or by adhering to certain requirements for adopting new curriculum based on state mandated standards and curriculum frameworks.

Research has tracked how some state and federal policies have influenced student and teacher outcomes. A great example of this is the Tennessee STAR experiment, where state-level administrators implemented a class-size reduction using a randomized control trial (Mosteller, 1995). The study showed the positive effects in some grades of class-size reduction. Other states, like California, used the Tennessee research to justify new class-size reduction policies, even with some researchers questioning the replicability of the Tennessee effects (Hanushek, 1999). In mathematics, a similar thing happened in 1997 when California's state-level math guidance encouraged school districts to teach Algebra 1 in 8th grade, under the old standards. As discussed in Chapter 1, California's expansion of Algebra in 8th grade was quite widespread. By 2013, 68% of districts across the state were teaching Algebra in 8th grade (McEachin et al., 2019). Yet as we learned earlier, the results of the state-level push for Algebra in 8th grade are mixed, and in some studies, like the study by Finklestein and colleagues (2012), only a small portion of students who have to repeat Algebra after 8th grade attain proficiency on the state exam in Algebra.

## LESSON LEARNED

Existing policies and practices may buoy or stymie attempts at detracking. Be sure to interpret whether the policies will support or hinder your efforts.

## Influential Community Groups

Community groups may voice support for or opposition against new policies. Community groups come in all shapes and sizes and can wield tremendous influence. Their support or opposition can make or break the development, passage, and maintenance of a math detracking policy.

For example, teachers' positions on such a policy may depend on whether they have experience and comfort working with students from diverse mathematical experiences. Some teachers may prefer math courses to be grouped by perceived ability in mathematics and may express opposition to a detracking policy. Others may be more motivated and open to a change to support the goals of equity, but will need coaching and professional development, or they may doubt their success and want to teach what is seen as an *easier* class. Because teachers' unions play an important role in policy development, and as the voice of the teachers they represent, they may influence any effort to change state-level policies related to the design of math courses across schools.

Likewise, community and family groups may coalesce around specific interests and may either support or oppose new policies. Sometimes new policies are developed as a result of advocacy from parent or community groups, and consequently they are very supportive and advocating to adopt the policy. Other times, the development of new policies can be questioned by communities and family groups based on their opposition during the phase when a policy is debated by the school board before being passed.

In SFUSD's case, the district's efforts to change their math course sequence had support from community groups like Coleman Advocates, who are focused on effective schools for all families and children, with a special focus on low-income families of color. Many math teachers in SFUSD were likewise in support of the mathematics policy. Some teachers had been trained in a pedagogical approach called *Complex Instruction* (Cohen & Lotan, 2014), which encourages heterogeneous groupings of students in classrooms for optimal learning. Many of those teachers trained in Complex Instruction reported having successful experiences in teaching heterogeneous classes. However, the SFUSD school board also heard concerns expressed by Parent Teacher Association (PTA) groups as well as support from parent groups like the African American Parent Advisory Council (AAPAC). The SFUSD leaders participated in community meetings and back-to-school nights to explain the new policy and hear both concerns and support for the policy from community and family groups.

Community groups are important groups to pay attention to in the educational context, especially if they are well organized. Teachers' unions are arguably the group having the longest history of being well organized. Cowen and Strunk's (2015) review of research on the effects of teachers' unions show their influence on everything from state and federal policy, to their association with changes in student outcomes. Other groups, like families, are also well organized, with some subgroups being more disenfranchised than others because of racism, sexism, and other struggles for power and status. For example, SFUSD historically has not had a strong reputation for engaging with its community of Black families. Only more recently has SFUSD leadership created an AAPAC that provides Black families with a formal body to communicate directly with the school board and district administrators.

## LESSON LEARNED

It is important to be aware of community groups that may influence the success of a new math policy and find ways to work productively toward common goals.

## Historical Events

The history of any community influences the cultural and social norms of its members. Consequently, history creates unwritten rules about what is or is not talked about, how people are treated, and who has the power and status across certain dynamics. Historical events can happen within schools and districts—like efforts to desegregate schools—or they can happen outside of schools—like the shooting of Dr. Martin Luther King Jr.—that still influence realities across contexts today. Historical events can create cultural and social norms and unwritten rules that may undermine or bolster new policies rather than leveraging norms to buoy a policy.

The important question to ask here is what events from history could collide with a potential change in policy? Are there certain historical events in a community that have developed norms or policies? One dramatic example would be how some schools that have experienced shootings, or threats of them, have extra security guards, require students to wear clear backpacks, or may even have extra curriculum to address the trauma or the social-emotional well-being of students to avoid future shootings.

What are historical events that could derail or support a change in math coursework? In looking at SFUSD as a case study, the historical events in the school district were the litigation related to student assignment. Two law-suits had been previously filed against the school district for policies that assigned students to certain schools using factors like race. Community groups like the NAACP and the Asian American Legal Foundation were involved in the lawsuits in hopes of changing SFUSD's student assignment policies. Consequently, SFUSD has a history of community groups engaging in the policy development process within the school board.

Probably the most pervasive sociocultural reality in most school and district communities is the reality of systemic racism and growing income inequality. Regarding racism, similar to many communities across the United States, the city of San Francisco and its schools were designed to support exclusion of certain racial groups through policies like school segregation and redlining in housing policy (Quinn, 2020). The impact of these racist systems still exists today across all educational systems given the substantial gaps in opportunity experienced by historically underserved students and families, like Black, Latinx, and low-income students.

> *Probably the most pervasive sociocultural reality in most school and district communities is the reality of systemic racism and growing income inequality.*

# LESSON LEARNED

For leaders and educators new to a district or school, it is important to learn about what local historical events may shape the school's, district's, and community's current norms and policies. Likewise, state and federal leaders may need to understand and factor in historical events that could influence their education legislation so that their policies do not get weighed down by litigation from local educational leaders or community groups.

## Questions to Consider for Your Context

▶ As you and your team design your goals and plans related to detracking math courses, it is important to think about the context in which those goals and plans will be implemented and achieved. Here we provide some practical questions you and your team can explore. These will help you think about your context to support your *premortem* as you anticipate what barriers or pitfalls may undermine your desired policy.

- **Leadership:** Who are the influential leaders in your community who will either support or oppose your policy? What could they say that would challenge the policy? What would convince them to support the policy?

- **Goals:** What are the goals of your classroom, school, or district? How do those goals relate to an effort to detrack math courses? How could those goals support or conflict with your efforts to create a detracking policy?

- **Legal decrees:** What are the policies created as a result of legal action? Do they have a connection back to math courses that could influence your effort to detrack math courses?

- **Influential community groups:** What groups are organized enough to create barriers or dissuade the public from supporting your math detracking policy? What marginalized community voices have not been heard?

- **Existing policies, practices, historical events:** What are existing policies, practices, and historical events in your classroom, school, district, or state that could undermine a detracking policy? What math curriculum has been adopted? Are teachers and schools enacting the curriculum? What classroom practices do you observe in middle school and high school? What are math placement policies (written or unwritten)? Who holds the power in those placement policies? Who holds the power in your school math teams? What positionality does your school district math department or staff have?

# Activity 2: Conducting a Premortem

▶ Materials: paper, pens, chart paper, markers

Time: 1 hour

Step 1: Identify a facilitator and notetaker. Gather a team of school and district personnel and interested community members who have the highest interest in developing a detracked math course policy. Describe to them more generally the detracked math policy your school system aspires to. Have them pretend it's been 2 to 3 years since they started implementing the policy and there is now evidence that the policy is being challenged by particular parts of your system.

Step 2: Have each person on the team write for 5 minutes on their own all the reasons they can think of that the policy would have failed.

Step 3: Have one person at a time share one reason they wrote on their own list. Have the notetaker write on the chart paper with markers for each of the reasons. Keep having each person share ideas one at a time until all the ideas people thought of are listed.

Step 4: Use the questions above at the end of this chapter to examine your list of reasons for failing. Do the questions help you add to the list of potential failures?

Step 5: Divide the list of reasons for failing into a two-column list. In one column, put the reasons for failure that you believe are within your school's or district's control. In the other column, list the reasons for failure that the team feels less confident about controlling. Use this list to help you develop or revise your current plan for detracking your math courses.

# CHAPTER 3

..............................

# DESIGNING A DETRACKED MATH COURSE POLICY

**T**wo district leaders and their research partner were meeting in the district office, thinking about how to implement the Common Core State Standards for secondary math (CCSS-M). They were having a deep discussion of how one mathematics course would progress to another different pathway in high school. At this time, the two district leaders, Angela Torres and Ho Nguyen (both authors on this book), were teachers on special assignment (TOSAs), working as math content specialists for SFUSD. Their main responsibilities were working with teacher leaders, coaching, and providing professional development for secondary teachers. When exploring the new "Common Core" standards, they were relying on their own expertise built from years of working as teachers and systems leaders. At one point in their discussion, they realized they had a million questions like, "How are other districts making changes to their practices and structures to prepare to implement these standards?" and "What does research evidence say about mathematics course progressions?" With all these questions, they realized they needed other types of expertise and evidence to inform their planning for implementing the new standards.

The Strategic Education Research Partnership (SERP) had been connected with SFUSD several years prior, linking researchers to practicing teachers and in particular working with some middle school teachers to find ways to get more students to be able to make sense of and solve word problems. This partnership connected the San Francisco math team to Harold Asturias, the director of the Center for Mathematics Excellence and Equity at Lawrence Hall of Science at UC Berkeley, and Phil Daro, one of the authors of the Common Core State Standards. This work then grew to be a partnership to support the district toward implementing the Common Core district wide, with both partners bringing a national perspective, expertise on math education, including research, knowledge of math reforms happening in other districts, experience with professional development, and their connections to other experts in this field.

*Part of the conversation involved trying to solve the dilemma that the new mathematics standards changed the way certain concepts shifted across grade levels. This conversation was happening during the Spring of 2013, when SFUSD was still working under the old 1997 California State Standards that named Algebra 1 as the 8th-grade course. The CCSS-M now had a full, rigorous Math 8 course that included almost all the linear algebra standards, plus transformational geometry and statistics. The high school CCSS-M algebra standards would now include more time to explore quadratic equations and functions and also included new standards on modeling with linear, quadratic, and exponential functions, as well as categorical and quantitative data topics. Both CCSS-M Math 8 and CCSS-M HS Algebra 1 courses were such full, rich, and rigorous courses that compressing content to include Algebra 1 in middle school seemed impossible and not in the best interest of students' learning.*

*As they were discussing, someone said, "I wish we could detrack." Harold responded, "Why not?" Given all the changes that the district leaders would need to make in implementing the Common Core Standards, this was the moment to leverage those changes in ways that would detrack to remove one of the obstacles that has caused large inequities. This moment sparked months of research, discussion, and deliberation about detracking, possible acceleration points, and course pathways. Phil and Harold had also been working through SERP with Oakland Unified School District just across the bay over similar issues, and they knew the SFUSD team shared similar visions for math equity. They brought the two groups of math district leaders together to write a position paper that captured everyone's best thinking. This journey meant that teams needed to reflect on their values and beliefs, know the landscape of the math programs in their districts, and figure out what was needed for their contexts. The paper explained the rationale for their policy and became the foundation for their detracked secondary course sequence policy.*

## QUESTIONS TO CONSIDER WHILE READING THIS CHAPTER

This chapter explores a few essential questions, organized to provide an understanding of how to design a detracked math course policy:

- **What are your equity beliefs and values about mathematics and students?**

- **What are your current math course pathways? What does success look like in these pathways?** How proportionate (or disproportionate) are your enrollment data in your highest math courses at each grade level span (elementary, middle, high) when disaggregated by race, gender, emerging multilingual learners, students with Individualized Education Plans, social economic status, and so on?

- **How does current and past research support detracking?** Has there been any research conducted in your school or district examining course pathways? If so, what does it say? How does that research compare to findings from national research?

- **What is the proposed alternative to your current course sequence?** How do your vision, values, or beliefs play out in your course sequence proposal? How does your new course proposal eliminate or minimize tracking while allowing for students to still reach AP math in 12th grade?

## DESIGNING A DETRACKED MATH COURSE POLICY

Designing a detracked math course policy is a complex task that requires research and a clear description of the *why*s behind the decisions that go into creating these course pathways. A team needs to be clear about needs in their system and what equitable outcomes for all students look like throughout the school years. Decisions on course options for students should be rooted in your equity-centered vision, show how your current course pathways are inequitable for students and are in need of a change through local data, and are connected to a variety of research in the field that give hopes of how the proposed change can yield more equitable outcomes. Additionally, your proposal must be packaged in a way that district administrators, teachers, families, and community members can understand the *why* behind the changes while allowing each to see how all students have rigorous pathways in front of them that they can find success in. This chapter outlines some proposed steps a school or district might take when developing a proposed math course policy.

## Step 1: Develop an Equity-Centered Belief Statement

The work to detrack a system is hard and difficult because it likely goes against a system that has been in place for decades. For any person or team to persevere through the work it takes to shift such systems, it is important to draw strength from your defining individual and collective values and beliefs, along with those of your allies. The first step to considering a detracking policy starts with understanding and articulating the values and beliefs that drive it. This in itself is a multistep process that includes

(a) self-reflection

(b) collective reflection

(c) searching for examples

(d) developing or revising a vision statement

These steps are not done linearly and can happen simultaneously, with each part reinforcing the others. For both self-reflection and collective reflections, here are some questions you need to ask that will support you to interrogate the inequities that you know exist:

- Is it acceptable that there are continuous patterns of low achievement for different groups of students?

- If not, what in your system produces that pattern of low achievement?

- What happens on a daily basis in classrooms that reproduces those results?

- In my (our) own teaching practice, did those same patterns exist?

- What do I (we) value as learning mathematics, and how have those values supported students to learn, or how have they reproduced the inequities?

- I (we) may believe that all students are capable of achieving, but do I (we) believe they *already* have mathematical strengths and knowledge?

Seeking out examples of classrooms where you see diverse students, especially Black, Latinx, and other underserved groups in mathematics, engage in rigorous, grade-level math is critical in pushing oneself and a team's belief of what is possible. These examples can come from classroom visits of your schools, videos of classrooms from your school or district, or even as a starting place, videos of classrooms you may obtain through conferences or math groups and/or organizations (such as https://www.insidemathematics.org/). Having common experiences observing or watching videos of diverse students and making the time to name mathematical strengths of students and reflecting on what was observed will help further individuals' or group's conception of what is possible. Other times, it may be that a single member of your team, or a small group within your team, may have been inspired by an article, a speaker, or a conference presentation. There needs to be space for those team members to add to the collective conversation, especially if those voices belong to an underrepresented group to keep pushing the boundaries of what is currently understood as equitable. From these experiences, you will be able to begin to develop a vision statement or reexamine an existing vision statement to make sure they capture your values and beliefs that support equity in mathematics.

In addition to your equity vision and mission statements, leaders, schools, or districts can take advantage of national position statements. NCSM: Leadership in Mathematics Education and TODOS: Mathematics for All (2016) have published a position statement called "Mathematics Education Through the Lens of Social Justice: Acknowledgement, Actions, and Accountability" that challenges each of us to "eradicate mathematics as a gatekeeper, engage in the sociopolitical turn of mathematics education, and elevate the professional learning of mathematics teachers and leaders with a dual focus on mathematics and social justice." This challenge includes detracking. The National Council of Teachers of Mathematics (2018) publication *Catalyzing Change in High School Mathematics: Initiating Critical Conversations* recommends that "[h]igh school mathematics should discontinue the practice of tracking teachers as well as the practice of tracking students into qualitatively different or dead-end course pathways" (p. 7)

Your vision statement, along with national ones, can be used as a critical lever to push for change, because more often than not, almost all our districts and schools have not achieved them. If we (our collective we) have not made significant progress toward equity, then there needs to be a change. For the SFUSD central office mathematics team and STEM department, it was critical to lean into racial inequity because that is embedded in the district goals and in each member of the central math team's personal belief system. Kendi (2019) defines racial inequity as "when two or more racial groups are not standing in approximately equal footing" (p. 18). He defines being anti-racist as "one who is supporting antiracist policy through their actions or expressing an antiracist idea" (p. 24). To not act and to go along with the status quo that continually produces inequities

> *To not act and to go along with the status quo that continually produces inequities means we reinforce those inequitable systems, whether we intend to or not (Kendi, 2019).*

means we reinforce those inequitable systems, whether we intend to or not (Kendi, 2019). The work to remove barriers to more advanced math courses for underrepresented students is anti-racist work. When the SFUSD central office mathematics team began this work, they were fortunate that SFUSD had these three goals as a district:

- Access and equity: Make social justice a reality by ensuring every student has access to high-quality teaching and learning.

- Student achievement: Create learning environments in all SFUSD schools that foster highly engaged and joyful learners and that support every student reaching their potential.

- Accountability: Keep district promises to students and families and enlist everyone in the community to join in doing so.

The team also realized that it did not have similar goals stated specifically for mathematics. There were certainly unwritten values and beliefs about mathematics and students, but for the work that they were about to do, they had to make time to put them in writing. From this process grew their four guiding principles, which are stated here, and are discussed in more detail in later chapters:

- All students can and should develop a belief that mathematics is sensible, worthwhile, and doable.

- All students are capable of making sense of mathematics in ways that are creative, interactive, and relevant.

- All students can and should engage in rigorous mathematics through rich, challenging tasks.

- Students' academic success in mathematics must not be predictable on the basis of race, ethnicity, gender, socioeconomic status, language, religion, sexual orientation, cultural affiliation, or special needs.

The development of these values eventually led to SFUSD's current vision statement for mathematics: *All students will make sense of rigorous mathematics in ways that are creative, interactive, and relevant in heterogeneous classrooms.* After further reflection, the SFUSD math team realized that the guiding principles and vision were founded on two premises: *All students are mathematically brilliant*, and *math is a web (not a ladder)* (Featherstone et al., 2011). The first premise holds each educator and the community accountable for looking for the strengths of students that we *know* exist. The second references the fact that math is expansive. There are all multiple connections and ways of thinking among mathematical topics (the web) and math is not learned in a strictly linear progression where one skill must be mastered before another is introduced. Recognizing this helps educators and the community create access points for each student by making mathematical connections to what they do know and using their strengths to build new knowledge rather than fixating on and limiting instruction to what are perceived to be gaps or weaknesses in previously taught content. All SFUSD's work moving forward were explicitly connected to these values and beliefs in the guiding principles, vision, and premises.

## LESSON LEARNED

It is not only important to know your values and beliefs about equity, but they also need to be written explicitly so they can be leveraged internally and externally.

## Step 2: Examine Data to Learn Where Inequities Exist

The next step in developing a new policy related to math course taking will be to look at the current state of your math course pathways using administrative data. It may be difficult to resist what Rochelle Gutiérrez (2008) calls the "gap gazing fetish" of comparing standardized achievement data (often test scores). In her article, "A Gap-gazing Fetish in Mathematics Education? Problematizing Research on the Achievement Gap," Gutiérrez writes

> At their most extreme, achievement-gap studies offer little more than a static picture of inequities in schools. Because these studies rely primarily upon one-time responses from teachers and students, they can capture neither the history nor the context of learning that has produced such outcomes. And, whereas researchers can highlight the variables most closely associated with the gap (e.g., income, family, background), those variables are often not reasonable levers for change in the mathematics education community. Moreover, the cross-sectional nature of most achievement-gap data analysis means that they fail to capture student gains or mobility. (p. 358)

She goes on to say that "[r]egardless of the form of the data, the theoretical lens used to view the achievement gap is what supports deficit thinking and negative narratives about students of color and working-class students" (Gutiérrez, 2008, pp. 358–359). To avoid or mitigate the deficit-ridden pitfalls and limitations of gap gazing over achievement data, it's necessary to look at a broader and richer array of systemic and institutional metrics and systems. This would include to explicitly question the entire system from an anti-racist lens that has produced such gaps. Gutierrez suggests that studies should focus on student advancement toward excellence and comparing achievement within racial and ethnic groups, with emphasis on growth over time. What this array looks like will be unique to your own context. For example, some data that can be considered include

- Math course enrollment, especially advanced math courses, disaggregated by race and/or ethnicity compared to proportionality of the school population

- Math course enrollment in advanced math courses compared within racial and/or ethnic groups (for high school, a 25% marker being the minimum for an equitable system can be used as the comparison since high school matriculation is usually 4 years and we would expect and want at least within 1 of those 4 years, a student would be taking an advanced math course such as AP Stats, AP Calc, Precalculus, etc.)

- Mathematics course grades and/or math grade point average (e.g., What GPA is considered excellent? How does the GPA of Black students in lower math tracks compare to that of Black students in higher math tracks? How has their GPA changed over time?)

- Math university and college eligibility completion (e.g., What is the rate of university eligibility of Latinx students in lower math tracks compared to Latinx students in higher math tracks? How do those compare to the average rates?)

- School or district performance assessment data (e.g., What does your performance assessment say about how students think and what they know? How has that learning progressed over the years, with excellence defined by the assessment rubric?)

When students within a racial group do not meet the standards for excellence repeatedly, then questions about the classroom, school, and district systems need to be asked and actions are needed to make changes.

The SFUSD school district mathematics team initially focused on enrollment data. They chose to collect data by visiting every high school, talking to department chairs, and reviewing course offerings in their middle and high schools. They had already made classroom visits to about 90% of high school math teachers' classes, when most if not all the district's high schools had tracked math courses. The team witnessed that enrollment of Black and Latinx students was most prevalent in the lower tracked math classes and that AP mathematics and fourth-year math classes had few and in some cases no Black or Latinx students. These observations uncovered the disproportionate enrollment in these classes when compared to the percentage of Black and Latinx students in the district.

When examining enrollment data for 8th-grade math for the middle school graduating classes of 2008–2010, the team saw that Black and Latinx students were *four* times more likely to be enrolled in General Math verses

(Pre Common Core) Algebra 1. This did not sit right with the team's values and beliefs about students and mathematics. And it went against the district and department's vision and mission. These data helped start the process of rationalizing a new mathematics policy.

## LESSON LEARNED

Be aware of and avoid participating in *gap gazing* by focusing on gaps in achievement in math through a deficit lens or blaming students for the gaps in achievement. This work must be done by looking truthfully at a variety of data and patterns that reflect your context. What story does the data tell about the equity or inequity in your system?

### Step 3: Review Research About Tracking and Detracking (and Other Inequities)

To add to the data you collect locally, you can also examine data from broader research about tracking and detracking in mathematics. Start by using a search engine like Google Scholar or ERIC by the Institute of Education Sciences (https://eric.ed.gov/) and search for "tracking," "detracking," and other search terms relevant to your context. Select research published in peer-reviewed journals or synthesis of research in literature reviews. Search for articles by Adam Gamoran, Jeannie Oakes, Pedro Noguera, Thurston Domina—some of the researchers who have published a lot of studies on detracking. Check out books in the library by Jeannie Oakes or Maika Watanabe. There is a lot of research out there, so reserve a day or two to review it. In Figure 3.1, we list some of the more practitioner-friendly summaries of research in commentaries and articles that are short, accessible, and mostly open access to everyone (without a paywall), as well as two books.

| AUTHOR, INSTITUTION | TYPE OF PUBLICATION | SUGGESTED RESEARCH COMMENTARY, ARTICLE, OR BOOK SUMMARIZING RESEARCH ON DETRACKING |
|---|---|---|
| Adam Gamoran, William T. Grant Foundation | Commentary summarizing research | Gamoran, A. (2021). In high school math, more instructional time helps, but the tracking dilemma remains. Commentary. *PNAS, 118*(29), 1–3. |
| Jeannie Oakes | Article | Oakes, J., & Wells, A. S. (1998). Detracking for high student achievement. *Educational Leadership, 55*(6), 38–41. |
| | Book | Oakes, J. (2005). *Keeping track: How school structures inequality.* Yale University Press. |
| Maika Watanabe | Book | Watanabe, M. (2011). *"Heterogenius" classrooms: Detracking math and science: A look at groupwork in action.* Teachers College Press. |
| Pedro Noguera | Journal article summarizing research | Rubin, B. C., & Noguera, P. A. (2004) Tracking detracking: Sorting through the dilemmas and possibilities of detracking in practice. *Equity & Excellence in Education, 37*(1), 92–101. |

Another way of accessing reliable research is to work with a research partner. In the case of SFUSD, the team relied on Harold Asturias from the Center for Mathematics Excellence and Equity at Lawrence Hall of Science at UC Berkeley as well as others working with SERP. These partners would help the team wade through the research and recommend certain specific research to read.

The SFUSD team found research by Carol Burris, Jeannie Oakes, and Jo Boaler to be useful in explaining both the challenges with a tracked math system as well as support for a detracked math system. In addition, Maika Watanabe's summary of research on tracking was particularly motivating because she laid out the reasons why tracking is problematic as district leaders worked to propose a detracked system (see Chapter 9 for further discussion on the use of research on detracking).

## LESSON LEARNED

When designing a policy, search for the research and literature on detracking and tracking to find similarities to your context and ideas for how to best support your own unique detracking efforts. Look nationally and globally for research done, including books that have compiled examples from other research studies, and look to national and local experts who can support you in this endeavor.

### Step 4: Designing an Alternative Course Sequence With Enough Detail

Once you have identified the patterns in your system based on what you've learned from your data and research, what will you do with it? Assuming that your current course sequence and math program is producing inequities, the next step is to design an alternative. Here are some questions to consider, in no particular order, when designing an alternative course sequence:

- Which would better serve your goal for change, an integrated course sequence or a traditional course sequence? What about your contexts supports one over the other?

- What are possible rigorous fourth-year math course options that support college and careers (and are not dead-end courses)?

- Given your current contexts (see Chapter 2), how much can you push the system? Can you detrack in grades K–10? Or just Grades 6–8? Or 6–12?

- How do you define rigor? And how would you ensure every course is rigorous?

- Do you want to provide acceleration options for those wanting to major in STEM fields? When should that happen?

- How do you prevent acceleration pathways from becoming tracks?

- Does your proposed sequence address the equity vision and goals of your district or school?

After reflecting on and articulating their equity vision statements, reviewing local data, the research, and studying the new mathematics standards, the central office mathematics team realized that the work to detrack their current system meant they needed to have an alternative secondary course sequence in their proposal. They define a course sequence to mean all the possible pathways that students can progress through their math courses from Grades 6 to 12. They borrow the National Council of Teachers of Mathematics' (NCTM) *Catalyzing Change* (2018) definition of pathway to mean a particular "course progression for a student through high school mathematics. Pathways can include tracks—fixed consequences of courses that are often determined in middle school or earlier" (p. 16).

When the math team considered SFUSD's goal of Access and Equity and their math department's Guiding Principles, they needed a new course sequence that removed barriers for each and every student. The sequence needed to better meet their math graduation requirement and university eligibility requirements promotes each student to take 4 years of mathematics (without repeating a course) and would provide access to AP Calculus, AP Statistics, and other advanced math courses for any students who wanted that option.

The SFUSD math team designed this new math course sequence by examining their interpretation of the evidence in light of their vision and goals (e.g., their interpretation of the Common Core State Standards), pinpointing the key decisions (integrated vs. traditional math), and defining relevant terms (e.g., rigor, acceleration). They started by asking themselves how the shift in the Common Core State Standards (CCSS) impacted the scope of content for each course and therefore how they needed to reconsider pathways because of those shifts. They also took into account the vision and guiding principles for mathematics that would help them achieve their goals of access and equity in mathematics. Based on this evidence and their vision and goals, they mapped out a design of a new course sequence that aimed to mimic their interpretation of the standards situated within the context of the SFUSD math program to date. Once they saw the draft math course sequence, they then asked themselves, "What does it mean to have acceleration options to AP Calculus without reinforcing tracking?" This led them to want to define terms like rigor and acceleration.

### Relying on the State-Level Decision to Adopt the CCSS:

A particular leverage point for SFUSD was that the CCSS shifted many of California's old algebra standards into Grade 8 while adding geometry and statistics standards that were not there previously, forming a full, rigorous Common Core Math 8 course. The Common Core also added more standards for the CCSS High School Algebra 1 course that includes additional standards to learn about quadratic functions, introduction to exponential growth, modeling with linear and nonlinear functions, and statistics (see Figure 3.2).

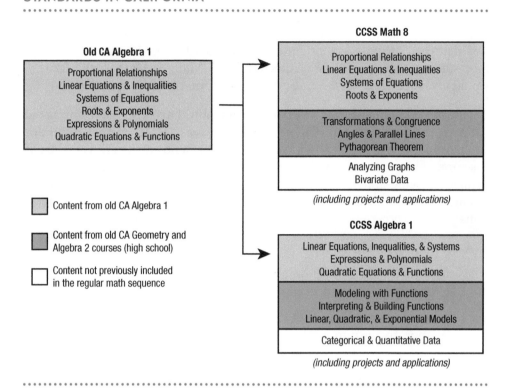

SFUSD's previous secondary course sequence had the old Algebra 1 course in 8th grade because the state of California defined those standards as 8th-grade standards. Before 2011, students in 8th grade were placed into two tracks: General Math or (old) Algebra 1. From 2012–2014, SFUSD worked toward the "Algebra for All" in Grade 8 movement in California. Both those course pathways effectively tracked their students, the first by sorting out students formally in 8th grade and the second by sorting students through high failure and repeat rates. Both of those impacts limited student options for higher math and were done disproportionately to their Black and Latinx students. The newly defined Common Core Math 8 standards allowed SFUSD to have a singular pathway in middle school: All students take CCSS Math 6, Math 7, and Math 8. This allowed the district to eliminate math segregation in the middle grades.

### Using the CCSS-M to Guide Decisions About Rigor and Acceleration

According to the Common Core, achieving rigor requires us to teach math in a way that balances students' conceptual understanding, their procedural skill and fluency, and their ability to apply what they know and are able to do to real-world, novel, problem-solving situations. The district math team also leaned on the idea that rigor meant going deeper, not faster, which was consistent with the team's understanding of the Common Core. They, alongside their teachers and research partners, embedded the Common Core Standards for Grades 6, 7, and 8 with the Standards for Mathematical Practices in their curriculum; they realized that each of those sets of grade-level standards would be rigorous for all their students. This could only happen if CCSS Math 6, Math 7, and Math 8 would give their students the *time* to experience that kind of rigor.

The proposed policy started with an agreement to have a singular untracked pathway in middle school. The next question was "How do they allow for acceleration for students who have STEM interests to reach AP Calculus as a senior, while not reinforcing tracking along the way?" Acceleration is defined as being able to reach AP Calculus by 12th grade without skipping essential content. These discussions lead the SFUSD team to solidify additional values and considerations:

- Acceleration should be delayed until high school because it
  a. supports the thoughtful progression of mathematics through middle school, some of the most important, foundational math students take
  b. supports younger students to develop strong math identities doing collaborative group work in heterogeneous classes
  c. provides the *time* needed for deep learning of essential concepts and to enact the Standards for Mathematical Practice, and to experience rigor
  d. equips high school students to make acceleration choices that match their long-term goals for college and career
- Mathematically, each CCSS Mathematics content standard is an essential building block for future learning; therefore no mathematics content or grade level can be skipped
- Compression (a course that contains more than a year's worth of standards for that grade level) as an acceleration option is better in later years because students are more mature and they are better able to commit to their choices
- Pathway choices for acceleration options should be made by students and their families only and not by singular measurements or recommendations by teachers or counselors—requirements that have historically become barriers for many students, especially our Black and Latinx students and all other students who have not been able to access more advanced math courses

After much debate and discussion using the processes above, the district math team came up with the new course sequence seen in Figure 3.3. In this course sequence, students have a singular pathway up to 10th grade after they take CCSS Geometry. The "Decision Points" are made by students and their families to accelerate by enrolling in an Algebra 2 + Precalculus compression course that would allow them to take AP Calculus as a 12th-grade student. Since then, SFUSD has added additional options for students to double up, or take two concurrent courses, in 9th (CCSS Algebra 1 and CCSS Geometry) or 10th grade (CCSS Geometry and CCSS Algebra 2), which will also allow students to take AP Math in twelfth grade. Each of SFUSD's high schools must offer at least one of these options. This is covered more in Chapter 10 where the authors share how SFUSD's pathway options have evolved over time.

**FIGURE 3.3  PROPOSED SECONDARY COURSE SEQUENCE FOR SFUSD**

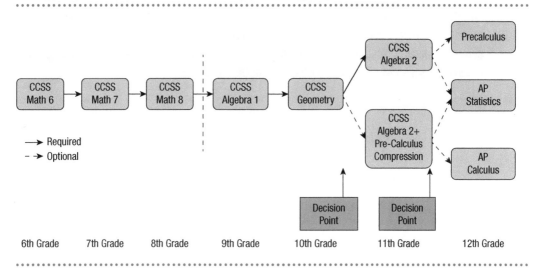

This course sequence is offered only as an example of how SFUSD came to a solution to the need for an alternative course sequence that would minimize tracking. There are many other possible course sequences and pathways that can eliminate or minimize tracking. This course sequence prevents segregating students until 10th grade, meets all their considerations, and answers many of their critical questions. SFUSD's new proposed sequence is in alignment with NCTM's *Catalyzing Change* statement:

> [A]cceleration should be along a single common shared pathway that provides each student with an opportunity to learn the same Essential Concepts . . . The divergence of pathways after students have learned the Essential Concepts in a shared pathway should be limited to a set of fluid course pathways that are open to all students. (p. 20)

*Catalyzing Change* suggests two sample high school course sequences called Pathway A that start with geometry and measurement standards first, followed by statistics and probability and Pathway B, which is an integrated course sequence. Another example is Oregon's High School Math Pathways 2 + 1 model (Oregon Department of Education, 2022). This course sequence has Algebra 1 for 9th grade, Geometry and Statistics for 10th grade, and then three pathways for 11th and 12th grades: Calculus/Precalculus Pathway, Data Science Pathway, and Quantitative Pathway.

## LESSON LEARNED

Design mathematics course pathways with intentionality to offer each student the foundational time to learn mathematics deeply based on the grade-level standards and the opportunity to take and succeed at a fourth-year math course that is setting them up for postsecondary success, including AP Math if they so choose, all while not allowing certain predictors, such as race, socioeconomic status, and home language, to determine which 4th-year math course a student chooses.

## Step 5: Making the Case: Proposing Changes to Your Course Sequence

Once you use the research to inform what could be the potential future state of your mathematics program, then it's time to propose any changes to the course sequence you think are necessary to achieve your school or district's goals. When advocating for a new policy in education, there is ideally a clear logic explaining the purpose behind the new policy as well as the reason a new policy was necessary. This is particularly important when there is potential for the policy to be controversial. What story does the data tell about the equity or inequity in your system? How will you connect your data, your story, and your new policy ideas to your district, state, and national equity statements to push for change? How will you share that story in all its complexity to the community, district leaders, teachers, and members of the Board of Education? This process is a delicate dance between the findings from the research (e.g., tracking creates barriers to math achievement for certain subgroups of students) and the realities of your school or district (e.g., some teachers like teaching in tracked classrooms). This is where your work from Chapter 2 to examine your policy context balanced with your values and beliefs that guide your math program will come in handy.

One approach to describing the change to your course sequence is by outlining your rationale for the change in a position paper. A position paper presents your proposed new policy backed up by specific evidence to support the rationale for the new policy. A strong position paper presents both sides of the argument and then uses the evidence to justify why one side is better than the other. (For more on how to write a position paper, see the activity at the end of this chapter.)

## LESSON LEARNED

It is important to lay out your case for a detracking policy into a clear and concise argument that can support others, outside of your math team, to understand the justifications behind the proposed changes. Community members, families, students, teachers, and administrators need to each be able to see *why* the changes were created (current inequities seen in data) and *what* these changes will look like so all students can reach a challenging fourth-year math course for their student, based on local data, research, and literature.

## THE ANATOMY OF A NEW MATH POLICY ON DETRACKING

Here the authors present the structure SFUSD leaders used to explain the proposed math course sequence policy, along with questions for you to consider in your own context.

- **Use data or evidence to explain why your school or district's math program needs to be improved.** Start by using existing data to explain the problem that this policy would be addressing. This information stems from your analysis of data examining the current or historical state of your math program. How is the math program currently designed, and what is going wrong with this design? What outcomes does your current program produce, and why is that a problem?

- **Show how your proposed policy is aligned with the goals you want to achieve in your math programming.** Maybe you want more students to access more math courses by the time they graduate, or you want more elementary students to achieve mastery of certain standards in math with a *consistent* marking. For SFUSD, they wanted to center equity to support each and every student to succeed in mathematics, with special attention to the students historically underserved who were traditionally not accessing higher-level math coursework in high school. What are your district's or school's vision, mission, or guiding principles and statements for equity in mathematics? What are your state's equity mission statements? How can you leverage these and those of national position statements?

- **Focus on the most significant details.** A successful policy is more about the *what* and less about the *how*. Show how your policy centers the features you believe are needed to make your math program work, but don't drown your audience in the details of implementation. For SFUSD, these included the move toward the CCSS in mathematics, which lead to a new course sequence. In the policy itself, they do not discuss the development of curriculum, support for instruction with professional development and coaching, or the community organizing and leadership needed to pass and maintain the policy. What are the variety of courses a student can take along the policy? Are choices for courses decided on by students and families? Can all students reach fourth-year advanced courses without needing to be in a particular path prior to high school?

- **Back the policy design features up with research and evidence**. Research and evidence can be used to justify a policy change, and it may also be used to explain a design feature of the new policy. For example, the SFUSD leaders relied heavily on their position paper on evidence from the CCSS in mathematics as a rationale for eliminating tracking through 10th-grade mathematics courses. How are you making the research and literature accessible to others to justify your course pathway options?

## Activity 3: Writing a Position Paper

▶ The leadership team at SFUSD used a position paper to explain the rationale for the detracked mathematics policy. It became a reference used throughout many meetings with individuals. While the position paper is not the policy itself, it shows a body of evidence supporting a certain viewpoint to rationalize a change or evolution in policy or practice and can be used as a foundation for crafting communication to different audiences like the school board, community members, and school sites. Here are the steps you can follow to write a position paper.

**Step 1**: Clearly explain the position or argument you are trying to make in a one-to-two sentence thesis statement. This thesis can either explain a challenge experienced in your school or district setting or a program that is showing promising evidence. For detracking, this could mean arguing that tracking influences your desired outcomes in a certain way or explaining how a new program, such as an effort to have heterogeneous math classrooms, can overcome challenging realities caused by detracking.

**Step 2**: Back up that thesis with multiple forms of evidence to support your argument. The SFUSD team relied heavily on research outside of their school district context, combined with data about student outcomes in mathematics within and across their schools.

**Step 3**: Explain a recommendation for changing a policy or practice in your school or district setting that would address your thesis. Now that you've made your case for change, it is time to describe a new policy or practice that either addresses the challenges presented or further supports the promising program.

# CHAPTER 4

........................................

# GAINING SUPPORT FOR A DETRACKED MATH COURSE POLICY

*W*ith poster paper and markers spread along the conference table, Superintendent Richard Carranza's Executive Leadership Cabinet sat together in teams of three and four, ready to do math *as part of a presentation from the Curriculum and Instruction Department. At the invitation of the head academic officer, STEM executive director Jim Ryan and math program administrator Lizzy Hull Barnes—one of the authors of this book—were making a case for employing a task-based curriculum as one way to move toward the mission of the district. A task-based curriculum focuses on using rich math tasks as part of a balanced approach to mathematics that includes conceptual understanding, problem solving, and procedural fluency. Rich tasks offer every student opportunities to engage in meaningful, rigorous mathematics, take time to solve, and lend themselves to collaboration and multiple perspectives.*

*The Common Core State Standards in Mathematics (CCSS-M) had made a compelling call for discourse-rich classrooms and math that provides for multiple means of representation. What better way to envision what is possible for students than to experience what is possible as a student? Today's work would be the 6th-grade task "Cups of Rice" (Illustrative Mathematics, 2016). The math leaders who were invited guests at this table established some very simple norms for doing math together and then asked the teams to engage in the task together and to make sure their whole team's ideas were recorded on the posters, with the guidance that we would end with a gallery review looking at one another's posters before debriefing as a class.*

*This direct experience of sitting in the chairs of student learners was not in the norm for this group of senior San Francisco Unified (SFUSD) leaders; in their current roles, they were more used to formal presentations that included slide decks and spreadsheets and talking points. After a beat of incredulous staring, this exchange happened:*

Superintendent Carranza (chuckling): "I knew you all were coming, but I didn't know you'd ask us to actually *do* math."

Another senior leader, a male former math teacher: "Say what you want, this is the most fun meeting I've attended at this table for years."

Superintendent Carranza: "You do know I taught social studies, right?"

*With the ice broken, the teams positioned themselves around the printed task and dutifully started uncapping markers and leaning over the awkward conference table, complying with this odd presentation. Soon, however, the greatest thing started to happen; in their own time and their own way, each of the three teams came upon language they actually had to discuss together out loud to understand.*

*This task asks students to make sense of fractional serving sizes and to describe this real-world situation with a visual model:*

Tonya and Chrissy are trying to understand the following story problem for $1 \sqrt{\frac{2}{3}}$:

One serving of rice is $\frac{2}{3}$ of a cup. I ate 1 cup of rice. How many servings of rice did I eat?

To solve the problem, Tonya and Chrissy draw a diagram divided into three equal pieces and shade two of those pieces.

Tonya says, "There is $1\frac{2}{3}$-cup serving of rice in 1 cup, and there is $\frac{1}{3}$ cup of rice left over, so the answer should be $1\frac{1}{3}$."

Chrissy says, "I heard someone say that the answer is $\frac{3}{2} = 1\frac{1}{2}$. Which answer is right?"

Is the answer $1\frac{1}{3}$ or $1\frac{1}{2}$? Explain your reasoning using the diagram.

*After some time spent on solving the problem, the group was brought back together to debrief the experience. In the debrief, one senior female leader directly addressed Ryan and Hull Barnes. As she gestured toward her group, she said, "I was trying to stay disengaged to see what you would do. When I was a student, I learned that girls could hide in math, and that was what we planned as a group. But then I had to explain the actual problem to him. Not the numbers or how to draw them or whatever, but the actual problem with the serving size."*

## QUESTIONS TO CONSIDER WHILE READING THIS CHAPTER

This chapter explores a few essential questions, organized to help you gain support for a new math policy focused on detracking:

- **Engaging educators in policy support:** Who are the educators or community members who could support your new detracking policy? For example, how might site-based educators, who are already leading equitable practices in their classrooms and schools, become advocates for larger systems change?

- **Engaging school and district leaders in policy support:** Who are the leaders in your school system who could support this new policy? How do district leaders learn about and adapt their own mathematical thinking to advocate and defend changing policies and practices?

- **Engaging community members in policy support:** How can educators work alongside families and communities to build collective advocacy? How can schools support different communities in seeing that detracking will support all learners, including their own community and their own child?

- **Engaging policy decision makers (e.g., school boards, mayors) in policy support:** Who are the decision makers in your context? How can you leverage work you are doing to support decision makers to make bold leadership moves?

## BUILDING A COALITION OF CHANGE AGENTS AMONG EDUCATORS

In every significant organizational change, there will be early adopters and leaders from the field who are already innovating with exactly the kind of leadership and collaboration any change effort needs. The work to detrack math courses starts with grassroots leadership from advocates in the field. These may be principals who are already using tools such as school scheduling to shift away from tracks and siloing, which you will read more about in Chapter 8. These may be classroom teachers who are using innovative practices like lab classrooms, lesson study, or peer-reciprocal observations to learn alongside each other. These same teachers might also be revising tasks to be more rigorous for a group of heterogenous students or making school-wide commitments about the mathematical habits of minds they are supporting collectively.

## Educators Currently Working Toward Your Vision

In the case of SFUSD, several high school mathematics departments were engaging in the deep pedagogical work of Complex Instruction (Cohen & Lotan, 1997). In classrooms where Complex Instruction was implemented, teachers were finding success with heterogeneous classrooms, creating student-centered and discourse-rich classrooms through the use of groupworthy tasks, participation structures, and the attention to status and accountability. The adjective *groupworthy* describes tasks worthy of collaboration, where students with diverse ways of approaching a task learn more deeply because they think differently and can build on each other's thinking.

---

### Groupworthy Task

This is a task designed to engage all students. In the book *Smarter Together! Collaboration and Equity in the Elementary Classroom*, Featherstone et al. (2011) ask three questions to support teachers in selecting or adapting tasks considered groupworthy. These questions are:

1. How can I make sure that this task will engage my students with big mathematical ideas?

2. How can I improve the chances that the context that the task proposed supports rather than distracts from the mathematical investigation?

3. Does the task allow for multiple points of access and more than one solution path?

---

We explore these ideas more in future chapters on curriculum and professional development. These educators who were already part of the Complex Instruction community were some of the strongest voices that spoke up at the Board of Education meeting when the team presented the detracking proposal, sharing examples of the successes in their heterogeneous classes. Many of these teachers also participated in SFUSD's curriculum development (see Chapter 5 for further discussion on curriculum development), bringing their lens of creating multidimensional tasks. The Complex Instruction program in SFUSD has since added middle grade and elementary school teams to this community, who bring their expertise and understanding of mathematics across the grade levels, even further expanding conversations.

While the SFUSD story included Complex Instruction as a shared experience that supported vision setting, your story may include other site-based reforms. Who are the educators who are already moving toward equitable practices in math? These early adopters are some of the first people you will want to collaborate with as you design a broader detracked system.

## LESSON LEARNED

Site-based educators are almost always leading from the middle; they are learner and teacher, follower and leader.

## Helping Educators Embrace the Vision

While you may already be working with many educators who support and are striving toward a vision of equitable practice in mathematics, you may also encounter teachers in your school, district, or state who believe that ability-based sorting best serves students. Their opinions may have been borne out of their experience, and they may not even be comfortable with the call to disrupt historical patterns of sorting students as a strategy for equity. Educators often join the profession because they are service-oriented, and they believe that education is a way to be of service to the world. It can be a hard truth to learn that there are structures in place in mathematics instruction—and in nearly all curricular areas and practices in education—that segregate or do harm in other ways.

There are several reasons why well-intended educators will have genuine questions about the purposes and practices of detracking:

1. Many people don't think of themselves as *math people*. Even in schools with the early adopters described earlier, there may still be educators who think they aren't good at math, or who believe the only way of being good at math is being fluent with algorithms and procedures.

2. Research might *tell* educators to change their thinking about themselves and their students, but reading other people's ideas does not usually move hearts and minds. These reluctant educators will grow their understanding as they see examples of students being successful in math classrooms, especially if these examples interrupt long-held beliefs about what it means to be successful in math or who can be successful in math.

3. Language around detracking can begin to sound inflexible and dogmatic; what will *not* support these educators is to talk down to them or correct their language as if this kind of deep transformation has a binary of right and wrong.

It takes bravery to challenge norms; if you and a small group of colleagues are the first ones to ask hard questions, you will most likely face pushback. That realization can lead to professional heartbreak and an urge or decision to leave the field, or that realization can become a source to tap into as you go deeper toward leading for change. You need to know *your* definition of equity in mathematics and why you care about making changes and challenging long-held practices.

> *You need to know your definition of equity in mathematics and why you care about making changes and challenging long-held practices.*

Site-based educators do not have the positional authority to make large-scale policy change, and yet have great influence with colleagues and students.

## LESSON LEARNED

Leverage the strengths, capacity, and resilience of site-based educators as you move toward bold reform.

## How to Be a Site-Based Educator Who Acts as a Change Agent

While you may be convinced on principle that detracking is a powerful strategy for equity in mathematics, it can be hard to know how to start in your role where you sit right now as you are reading this book. Here are some suggestions for site-based educators looking to get started:

- If you are a teacher, start by joining your School Site Council (SSC) or other site-governing body and invite other colleagues to join with you.

- If you are a site administrator, invite equity-minded educators in your building to join your site structures such as your school's governing body. Consider joining districtwide leadership, governing, and funding conversations yourself.

No matter whether you are a teacher or a site administrator, you can

- Create a book club or study group together around this book or other resources that discuss equity in mathematics or start a schoolwide practice of doing math tasks together

- Visit classrooms together, including at different grade levels and subjects

- Look for professional learning opportunities designed for heterogeneous classrooms

- Connect with labor partners such as leaders in the teachers' union or other community groups if any recommended changes to curriculum or policy will impact their working conditions or will influence the goals of their organization

## BUILDING A COALITION OF CHANGE AGENTS IN YOUR SCHOOL SYSTEM LEADERS

With significant change, senior leaders must be ready to guide the community "from the front of the room" while shaping the policy landscape. While these leaders must advocate, they may not have a background in mathematics and the research associated with tracking. Even senior leaders who may have taught math may not be aware of the recent pedagogical shifts in developing deep conceptual understanding associated with the reform movements in mathematics education. In this section of the chapter, we share ways to support senior leaders who will now be in a position both to advocate for and defend a policy. What are some ways to bring them into conversations about equity in mathematics and to equip them with the strategies and capacity to lead?

The vignette at the beginning of this chapter, albeit a narrative snapshot of one meeting, is one of many examples of how the math leaders in SFUSD supported the senior leaders to create a vision of what is possible in mathematics. While the senior leaders at the time were ready to move toward bold

*Through no fault of their own, many senior leaders think of math as speed and answer getting, similar to how they may have been taught.*

equity-based reforms, many of them did not have a background in mathematics education. Through no fault of their own, many senior leaders think of math as speed and answer getting, similar to how they may have been taught. You may hear senior leaders say things like, "I'm just not a math person," including those who may be leading from a place of equity.

In SFUSD, the team did not need to convince leaders that detracking was good work. They did need to provide talking points about the new arrangement of the CCSS-M—especially the arrangement of the standards in the domain of algebra. They also needed to support a narrative that included both research and math-specific localized data. As you read in Chapter 2, SFUSD's course sequence in math was not a stand-alone example of commitment to equitable outcomes. This commitment to equity work was true across structures and across content areas, for example including Ethnic Studies courses as a graduation requirement. As you now read this section of this chapter, consider for yourself what existing structures you can leverage. Are senior leaders in your context moving toward other bold reforms? If they are, how can you leverage that, and what can you learn about what is already happening? If this is a first attempt at shifting policy to better serve more students, why is detracking the compelling work you are taking up together as a system?

## How to Engage Senior Leaders in Detracking Work

While you may be working with a senior leader who already knows that detracking is a powerful strategy for equity in mathematics, it can be hard to support that person or that group of people to ramp up on classroom practices and a research base they may not themselves yet know. If you are a senior leader reading this chapter, these suggestions are also here for you:

- Start by understanding how decisions are made within the larger system around you.

- Meet with the most senior leader you have access to, and talk with them about their vision for equity, including but not limited to equity in math. Ask them what would most support them to clarify or develop that vision for mathematics. Do they need to understand research or local data from your context?

- As a vision-setting activity, do math together, modeling some of the facilitation moves you want to see in your classrooms.

- At each new juncture in your ongoing policy work, provide senior leaders with specific and clear talking points. Try hard to keep educational jargon out of these talking points.

- Share the most relevant and current research and position papers. Because almost every senior leader you will meet with has done significant coursework and developed years of practice in schools before being appointed to their current role, they will likely find research to be a meaningful and relevant entry point. They may also not be aware that because of such research every credible professional mathematics organization now supports detracking. Here are some resources you can share if senior leaders are not already aware of them:
  - Jeannie Oakes's decades of work as a leader talking about tracking as a class-based and race-based system may be familiar to them from coursework but is a good reminder (Oakes, 2005; Oakes et al., 1990; Oakes et al., 1997)
  - Professional math organizations have now made explicit calls to end tracking as one way to design mathematics for equity. These include the *Catalyzing Change* series (NCTM, 2018) and Leadership in Mathematics Education's position paper called "A Call to Detracking Mathematics" (NCSM, 2020)
  - Adam Gamoran's summary of research about tracking in a commentary to the National Academies of Sciences (Gamoran, 2021)

This work can only take root when senior leaders step up and lead with their positional authority, but they must have the research backing and support to do so.

## LESSON LEARNED

Support senior leaders to understand the work and why they should care about detracking.

## BUILDING A COALITION OF CHANGE AGENTS AMONG YOUR COMMUNITY LEADERS

Whenever we support policy shifts in schools, we must design *with* families, and not *to* families. Bryk (2010) describes how the quality of the ties between families, communities, and the school are directly linked to student motivation and participation in school. Henderson and colleagues (2007) in their book *Beyond the Bake Sale* explain how educators and leaders in education may collaboratively build relationships with families as a means to support the success of students socially and also academically. Therefore, if everything we are doing is to better serve the young people in our buildings, their identities, their development of content, and their growth as humans, then we must also understand that a family will always hold that far more deeply than we.

As you engage communities, remember to seek input and guidance from families and communities whose voices are not often heard. Who are the community partners who work alongside and advocate for these communities and advocate for social justice? In SFUSD's work, voices of support included members of the local NAACP, the African American Parent Advisory Council (AAPAC), Coleman Advocates for Children and Youth, and Chinese for Affirmative Action. Consider what voices you most need to seek out and learn alongside. As always, bring cultural humility to these conversations, especially if you are working across differences, remembering that most community-based organizations grounded in social justice have developed relationships and celebrate histories that go far beyond your work to detrack mathematics.

Engaging with some families about detracking may be especially challenging considering the pervasive public perception that detracking means *taking something away* from students. The most vocal opponents may be communities and families who have found success in schools with tracks, who feel that lifting all communities is conflated with diminishing rigor or lowering expectations. In this section, we share some of the resources SFUSD developed and some of the kinds of conversations they had, including those that went well and those that did not.

> *Engaging with some families about detracking may be especially challenging considering the pervasive public perception that detracking means taking something away from students.*

Early in SFUSD's implementation of the Common Core, the math leadership team participated in many family math nights, ideally presenting with teacher leaders or the site leader wherever possible. When the team hosts or supports events for families, they always do math together to help families experience best practices associated with math reforms such as the Common Core. This can include a math talk, which allows them to show what it means to develop flexibility putting together and pulling apart numbers, versus paper and pencil calculations such as a standard algorithm. They also share practical inputs that support this kind of shift, such as how to support their children with homework without teaching algorithms. In the case of SFUSD, every unit in the core curriculum also has a family letter, sharing the big ideas of the unit, what homework may look like, and some ways to help at home; many curricula feature family-facing resources such as this. Below is an excerpt from an SFUSD 3rd-grade unit on 2-D figures (SFUSD.edu, 2022):

## Helping Your Child With Homework

The Standards for Mathematical Practice describe the ways students behave as they learn math. While the mathematics content changes from grade to grade, these standards are the same for kindergarten through high school. Mathematical Practice Standard 8 says: ***Look for and express regularity in repeated reasoning***. Much of what 3rd graders do to make sense of the base ten number system falls within this standard.

One great example of this is when we practice with a series of problems using multiples of 10. Students see patterns in multiples of 10 and how the digit zero holds place value. Students recognize certain features of our number system and start to generalize things that will always be true.

$$6 \times 1 = 6 \qquad 6 \times 10 = 60 \qquad 6 \times 100 = 600$$
$$60 \times 1 = 60 \qquad 60 \times 10 = 600$$

You can help your child make generalizations by asking them what patterns they see. These are some questions and prompts that will help students make generalizations:

- What shortcut can you think of that will always work for these kinds of problems? Why does it work?

- What pattern(s) do you see? Can you make a rule or generalization?

Specific to detracking, the math leadership team also developed a set of resources to support families' understanding of the shifts in standards and curriculum. This included a Frequently Asked Questions or FAQ, which is a living document and continually updated, a family-facing fact sheet about research on detracking (see Figure 4.1), and a pamphlet for incoming ninth-grade families that shows visual representations of all the ways to accelerate learning of mathematics content in high school.

 **SFUSD** SAN FRANCISCO PUBLIC SCHOOLS

**Quick Facts About Math & Tracking**

**Q: Will taking away tracks in math (e.g. honors track math courses) help all students achieve at higher rates?**

A: *Studies demonstrate the positive impact of math coursework sequences that put all students through the same courses rather than tracking students based on their perceived ability.*

***Researchers from Columbia University found the probability of completing advanced math courses and math achievement increased in all groups when middle school students were enrolled in mixed-ability math courses.[1]***

Figure 1: **Increase in % of students participating in de-tracked middle school math courses that took courses beyond Algebra 2 in high school**

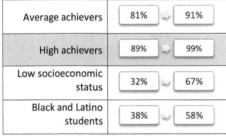

| | | |
|---|---|---|
| Average achievers | 81% | 91% |
| High achievers | 89% | 99% |
| Low socioeconomic status | 32% | 67% |
| Black and Latino students | 38% | 58% |

From: Burris, Heubert, and Levin (2006)

**More high-achieving middle school students in these mixed ability courses took the AP calculus exam and scored higher than students in tracked courses.**

This study of six middle school math classes in New York found that students' probability of completing advanced math courses beyond Algebra 2 in high school increased across all groups, including high-achieving students. Also, the average scores on achievement tests for high-achieving students who learned in math courses without tracks, i.e. heterogeneously grouped, were not significantly different than high-achieving students' scores in tracked math courses.

***Researchers from Stanford University and Kings College in London found all middle school students performed below their potential when in tracked math courses, both in high tracks and low tracks.[2]***

This study of over 1,000 students in London schools examined students' perceptions of going from mixed ability to tracked math courses during middle school. The results suggest that all students were negatively affected by the tracked math courses whether they were in the low track or high tracks.

**Students in higher tracks in math were disadvantaged by fast-paced lessons and pressure to succeed.**

[1] Burris, C.C., Heubert, J.P, and Levin, H.M (2006). Accelerating mathematics achievement using heterogeneous grouping. *American Educational Research Journal, 43*(1), 137–154.
[2] Boaler, J., Wiliam, D. and Brown, M. 2000). Students, experiences of ability grouping—disaffection, polarisation, and the struction of failure. *British Educational Research Journal, 26*(5), 631–648.

There are a number of things the team wishes they had done better and would recommend to others. You may find that some of these ideas are within your influence, while others may be beyond your positional authority.

1. Early on, the team was focused on removing barriers and providing access to advanced coursework but did not do enough to help families understand the beneficial impact this could have on each and every student's depth of learning and their future opportunities.

2. The team conducted many family math nights early on, as well as other kinds of public events such as district-level advocacy groups, but there was not enough consistency in how these happened year after year. Continuing these meetings or creating a module that site leaders could replicate and sustain might have helped families see how their children were top of mind.

3. Senior leaders took ownership of the policy through public statements, and high-profile leaders in the math community wrote opinion pieces for media outlets, but had the team themselves offered more opinion pieces, blogs, and podcasts about the detracking policy, families and communities would have been better informed.

4. Not partnering with more postsecondary institutions was an opportunity missed. While several members of the math team sat on intersegmental bodies that connect the PK–12 systems to postsecondary, both locally and at the state level, more conversations about alignment are needed in general across PK–16.

5. Another opportunity missed was not lifting the voices of students frequently. Early on, partners of the math team did conduct student interviews and share some of those valuable insights. This is another opportunity the team wishes they had been able to sustain.

## How to Engage Families in Detracking Work

Think about the families and communities who are experiencing your math policies and curriculum. How might you partner with them to not only better understand shifts in math teaching and learning but also get input and feedback? If you want to explore these concepts, there are many places to begin:

- Learn about the family advocacy groups in your context, both those that are formal, such as a district English Learner Advisory Council (ELAC), and those that are informally affiliated with schools, such as faith-based organizations that may support events such as after-school tutoring. What are the culturally humble ways to engage any given community? Introduce yourself and conduct empathy interviews or listening sessions. What do families need and want for their students? What strengths does each community bring?

- Put together early examples of family-facing resources. Ensure that they are translated into all the most frequent languages for your community. Try hard to keep educational jargon out of these resources.

- Look for early successes with outcomes as a result of your detracking, and share them. For example, if you change a policy that impacts middle school students, you will not see the downstream effect for at least 4 years. What are points of celebration along the way? This could be anything from grades to shifts in performance assessments to interviews with students.

Remember that every family is advocating for their child and wants to do anything that will best serve them. Come into any community space with a listening stance and empathy.

## LESSON LEARNED

Listen and practice humility with families, who are and always will be the first advocates for their children. Even if people may disagree with you, remember that they are coming from a place of wanting what is best for their child.

## BUILDING A COALITION OF CHANGE AGENTS IN YOUR POLICY MAKERS

At some juncture, you must formally present your ideas and your implementation design to policy makers or the people who make the decision about what policies to enact in your school or district. In the case of a school district or at the state level, this is likely a Board of Education, a mayor's office, city supervisors, or state congressional leaders. In the case of school board members, they may not be educators themselves and may have their own complex identities with how they or their families have experienced schools and mathematics. In this section of this chapter, we help you think through the kinds of things you might present to decision makers like school board members or other policy makers and how you might support these policy makers toward bold change.

One of the most striking parts of SFUSD's story was how complex the mathematical identities of those who were on the board were when the policy was passed. On the night in February 2014 when the team presented the proposed policy, after a series of other conversations, some board members described themselves or their own children disengaging with math at some point in middle school or high school and some said they were not *math people*. In each section of this chapter, we have offered practical suggestions for you to try and some of the lessons we have learned. All these opportunities, such as doing math with district administrators or offering jargon-free resources to families, will also support decision makers. The big difference, however, is that policy makers can ultimately make different recommendations, and these are the most public of spaces when advocates are needed to make statements on the record.

## How to Engage Policy Makers in Detracking Work

Think about those with greater decisional authority than you. How are decisions made within the larger system around you? Who are the policy makers making decisions about your systems-level policies? If you want to explore these concepts, here are a few places to start:

- Learn about the governance rules for the formal decision-making body in your school or district context. There are normally formalized structures for engaging with elected officials or other decision makers. Is there someone you know who can help you understand and navigate that?

- Engage in ongoing dialog with governing boards, to the extent that you can, with updates on your work in mathematics and specifically moving toward a detracked system. Policy makers want to feel engaged in not just the process of passing a policy but in maintaining accountability about the effects of the policy.

- Be real about what is going well and what you need to improve with your work to detrack math courses. In Chapter 10, we talk about the process of monitoring the policy. It's important to share the results of that monitoring and work to continually improve the policy with policy makers. You are engaging in what will likely be a multi-years' initiative, while those with decisional authority—the individual school board members and other elected officials—may change because of elections or assuming new roles.

## LESSON LEARNED

Be mindful that changes in governing boards can manifest huge shifts in your policy landscape. Be ready to start at the beginning as often as needed and remember that all systems change will have forward movement and backward movement. You will need to show up with resilience.

## Questions to Consider as You Work to Gain Support for Your Math Detracking Policy

▶ Here is a high-level summary and reminder of the questions we have asked in this chapter and the many questions to think through as you engage communities across roles.

- Who are the community members who are most supportive of your policy change to detrack math courses? Why do they support detracking math courses? What has been their personal as well as professional connections to math programming?

- Who are the community members who are least supportive? Why do they oppose detracking math courses? What are their personal as well as professional connections to math programming?

- What is your strategy for communicating with and engaging community members in various roles (students, families, teachers, school leaders, district leaders, policy makers) about the detracking policy? How can you make that communication ongoing rather than one-time only?

- Who is a team who will work on gaining support for the policy? For SFUSD, it was the math team within the curriculum and instruction department; this may look different across contexts.

## Activity 4: Build Support for the Policy by Doing Math Activities

▶ Engage instructional leaders and community members in math activities so they can envision what is possible in heterogeneous classrooms. There are many math activities you can do and many communities you will want to engage with. Here is one such example from the SFUSD story, working with secondary counselors.

In 2015, early in the detracking policy's implementation, the SFUSD math leadership team hosted a professional learning experience with middle and high school counselors, knowing that counselors are some of the most important people on the ground regarding providing advice and access to courses. While they shared some high-level descriptors and data, the bulk of the time was spent doing a Formative Assessment Lesson from Silicon Valley Mathematics Initiative or the Mathematics Assessment Project (2022), where the counselors were matching visual models to written descriptions to equations and justifying their ideas while making sense of the ideas of others. You can find these Formative Assessment Lessons freely at https://www.map.mathshell.org/. Finally, the team connected the math the counselors had done to the newly revised administrative regulations in their counselors' manuals.

These counselors, from many different educational backgrounds, were experiencing the vision of a math classroom where all students can be successful. They could now envision how any student placed in a heterogeneous class can be successful. During the card sort, the SFUSD math team asked the counselors to use these norms for heterogeneous classrooms:

**Groupwork Norms**

- All cards in the middle.

- Everyone touches the cards.

- Explain your thinking aloud.

- Ask questions to clarify and justify (how, why, what did you mean by . . .).

The team knew that this type of activity has many points of access so that each participant could find their way in and experience success, as was confirmed by feedback at the end.

Clearly, one math task—such as the Formative Assessment Lesson we did with these counselors—cannot drastically change a person's entire life experience let alone an entire system. And yet inspiring an equity-based vision of what is possible is crucial as you engage with various members of your communities. Doing math together, including doing math outside of traditional professional development spaces with those who are not math teachers, is a powerful starting place as you work toward a vision of detracked math classrooms.

Source: iStock.com/**FatCamera**

# PART 2

# IMPLEMENTING A POLICY OF DETRACKED MATH COURSES

# CHAPTER 5

........................................

# MATH CURRICULUM TO SUPPORT HETEROGENEOUS CLASSROOMS

I n 2012, *a curriculum development team of three teachers sat together to think about* groupworthy *tasks for 5th graders learning to read, write, and compare decimals to the thousandths.*[1] *This 5th-grade team from the San Francisco Unified School District (SFUSD) knew there were plenty of curricula full of procedural worksheets where students move decimals around to change the magnitude of a number, or pull numbers apart as a routine activity. Clearly,* moving a decimal *on a worksheet is not a big idea in 5th-grade math. It is a big idea that in the base-ten system, the value to the left of any number is ten times the value of a digit to the right. Students across their elementary careers need to grow a flexible understanding of this base numeration system and start to formalize zeros as placeholders and decimals as a mathematical way of noting fractions with denominators in multiples of ten. Yes, 5th graders need to pull those numbers apart, but for what understanding? Fifth graders need to compare numbers to the thousandths, but in what context? How would this development team work with this idea, without replicating all the fill-in-the-blank worksheets they had seen? Where is there a real-world math problem where students grapple with decimals to the thousandths?*

*At the time that this curriculum development team was grappling with this question, the San Francisco Giants had just won the 2010 World Series, and they were enjoying another compelling season in 2012 that inspired many of the students in SFUSD. Baseball statistics are a real-world example of decimals to the thousandths, with mathematical operations that can grow into high school math. Likewise, the Common Core State Standards for Mathematics had recently been adopted statewide, so the curriculum kept this real-world context at the forefront of their work to meet the call of the Common Core to create relevant mathematical situations. The real-world*

---

[1] CCSS.MATH.CONTENT.5.NBT.A.3

*connection helped them think about how to design a task to be groupworthy that could be authentic to the students' experiences at the time. The curriculum development team used the definition of a groupworthy task described in Chapter 4 to guide their design. The teacher developers foregrounded students' love of the Giants in tasks where students were comparing and ordering decimals. Here was an authentic reason to play with the decimals and to compare players' stats across different teams and different seasons.*

## QUESTIONS TO CONSIDER WHILE READING THIS CHAPTER

This chapter explores a few essential questions, organized to provide an understanding of the role of curriculum in detracking efforts. It will help you

1. **Get clear on what the word *curriculum* means to you as an individual educator and to you within the collective of your school or district:** Is the curriculum a set of standards? Or is curriculum pedagogical guidance? Is curriculum an adopted textbook? Does the curriculum include physical manipulatives or technology? What is the relationship (if any) between your curriculum and the sorting of students into different courses? In this chapter, we describe how the SFUSD team defined curriculum within our work; this is an important question for you to think through as well.

2. **Examine how your curriculum relates to design features in detracked math courses:** Is your curriculum aligned to your core beliefs about teaching and learning math in heterogeneous classrooms? How does the

curriculum offer multiple means of expression and representation for all learners? How does the curriculum support students to interact with each other as they make sense of the math? How does the curriculum support a discourse rich math classroom?

3. **Review the needs of students and teachers when using your math curriculum:** How are you revisiting and reflecting on your curriculum and its match to the needs of your students? How much agency do teams of teachers have to calibrate around and enact powerful curriculum in meaningful ways? What agency do educators have in your context to adapt and revise curriculum to match their students and their classrooms? What strategies are called out in the curriculum you are adopting or adapting? If strategies are not highlighted, how will your communities of teachers plan in ways that support peer-to-peer interaction? What are the authentic partnership bridges between educators in the classroom and those who are positioned to make decisions about curriculum?

## THE IMPORTANT ROLE OF CURRICULUM

When states adopted the CCSS-M or revised their standards in similar ways, many educators across the field saw an opportunity to redefine what it means to be a mathematically *smart* student. The Standards for Mathematical Practice—some form of which are embedded in every state's standards as practice or process standards—highlighted reasoning, precision, and mathematical patterning and also placed a high premium on behaviors such as constructing viable arguments and critiquing the reasoning of others. For heterogeneous classrooms to evolve as powerful places where students learn and thrive, the math in front of those students has to be math worth collaborating on. Tasks that are complex and that offer multiple solution pathways can provide an authentic reason for students to listen in and ask each other questions. Multiple representations provide authentic reasons for students to look for connections across different work and to make the kinds of connections that build conceptual understanding as well as procedural fluency.

Tasks that allow for multiple means of engagement, representation, and action and expression also align philosophically with Universal Design for Learning (https://udlguidelines.cast.org/), which is an approach to learning in which educational experiences are designed for all learners, as opposed to being designed with one *typical* user in mind and then retrofitted through differentiation for *other* groups of students. In essence, math in classrooms aligned to modernized state standards in mathematics, whether they are the Common Core or your own state's updated versions, requires the math of the social world, and not the math of rigid and procedurally focused textbooks.

Curriculum sits at the heart of the relationship between a student and teacher and is therefore integral to any policy or practice change in mathematics instruction. Think about City and colleagues' explanation of the "instructional core." City and colleagues explain the instructional core as the interaction between the teacher, student, and content. They argue the instructional core is where change needs to happen if school and district teachers and leaders ultimately want to support access, equity, and achievement for diverse student learners

> *Curriculum sits at the heart of the relationship between a student and teacher and is therefore integral to any policy or practice change in mathematics instruction.*

(City et al., 2009). In essence, the curriculum acts as the content, which is the first of the three conditions we think of as undergirding mathematics instruction for diverse learners.

The SFUSD Math Team has been growing and evolving their understanding of curriculum over many years. The SFUSD team developed their curriculum before curricula aligned to the new standards were available, which was both necessary and powerful work for them, and not what they would recommend to others being that there are now strong curricular resources available to you. For the purposes of this chapter, here are the early agreements the team made about math curriculum that would support heterogenous classrooms:

1. The SFUSD Math Core Curriculum would be a task-based curriculum with multiple embedded opportunities for formative assessment.

2. Pedagogical guidance for instruction foregrounded peer-to-peer discourse.

3. Lesson design gave the most time for student inquiry and exploration in collaborative groups.

4. The emphasis was on Tier 1 instruction.

5. Tasks have multiple entry points and are meant to provide every student with access and rigor.

The emphasis in number five is crucial: Curriculum is *meant to provide every student with access and rigor*. A curriculum designed for a heterogenous classroom would not provide rigor for some students (*high* or *fast* students) and access for others (*low* or *slow* students). Every student should have access, and every student should be challenged. When students are working with accessible and rigorous curriculum, they have a reason to collaborate. This idea connects naturally to the SFUSD Math premise that *math is a web (not a ladder)*, meaning that there are many ways to do and understand mathematics (Featherstone et al., 2011), discussed in more detail in Chapters 3 and 6.

## CONNECTING CURRICULUM TO YOUR BELIEFS ABOUT TEACHING AND LEARNING MATHEMATICS

A curriculum that is task-based is meant to provide every student with access and rigor. What this means within the conversation on detracking is that students must have powerful mathematics that is worth collaborating on

(Foster, 2019). If students work with curricula that favors speed and answer getting over sensemaking, then two things will drop away: First, the students will have no need to collaborate or to build on the thinking of their peers; and second, the beauty and complexity of mathematics as a web will no longer provide multiple access points and many ways of representing mathematical ideas.

Phil Daro, one of the lead authors of the Common Core, argues that we should see our students as problem solvers, not answer getters (SERP Media, 2014). Our students know if a teacher, school, or a system as a whole values their thinking. They also know whether the only way of demonstrating *being smart* in an educator's eyes is answer getting in isolation. In a classroom that values speed and answer getting, the *smartest* student is the fastest student, who computes an answer correctly and has no reason whatsoever to collaborate. This also sets up a competitive space where everyone is or is not *the smartest*. A performative and competitive culture in math flattens the beauty and multidimensional nature of mathematics, by treating it not as a discipline, but as a series of steps to be rehearsed and replicated for an artificial marker such as a chapter test. It creates hierarchy within a math classroom and exacerbates inequities. We are not saying that the answers don't matter, because they do. And yet many people who self-describe as *not good at math* or *hating math* are describing feeling unsuccessful in a rigid, prescribed, performative model of mathematics.

> A performative and competitive culture in math flattens the beauty and multidimensional nature of mathematics, by treating it not as a discipline, but as a series of steps to be rehearsed and replicated for an artificial marker such as a chapter test.

In addition to groupworthy tasks, problem solving, and collaboration, another significant idea in the Common Core and other state standards is the focus on conceptual understanding first so that opportunities for application and procedural fluency connect regarding sense making. In the National Council of Teachers of Mathematics' Six Principles for School Mathematics (2000), which predates the CCSS-M, the authors state that curriculum must be built conceptually. They state that a curriculum "is more than a collection of activities; it must be coherent, focused on important mathematics, and well articulated across the grades." As these ideas build, so will a student's understanding of the expansiveness and interconnectedness of mathematics.

In keeping with these long-held beliefs about balanced math curriculum, when the Common Core was introduced it signaled an important shift that stretched many people's ideas about mathematical proficiency. The Common Core now defines rigor as a balance of concept, procedure, and application and pushes directly against moving faster through topics: *Rigor refers to deep, authentic command of mathematical concepts, not making math harder or introducing topics at earlier grades* (Common Core State Standard Initiative, 2021a). Similarly, the Progressions Documents (see these documents from Achieve the

Core, 2013) provide important narratives of how math articulates within the Common Core, across the grades, within any domain of mathematics.

One framework that explains the behaviors and practices of mathematics is the Teaching for Robust Understanding (TRU) framework. Developed by UC Berkeley professor Alan Schoenfeld, the TRU framework suggests there are five dimensions of powerful classrooms that define the components of classrooms with robust instruction. The TRU (n.d.) framework says the dimensions are:

- The Content: the richer the content, the more students are likely to know

- Cognitive Demand: what opportunities are there for students to engage in productive struggle

- Equitable Access to Content: every student has the opportunities to engage in the core content

- Agency, Ownership, and Identity: students think about themselves as having a strong disciplinary identity

- Formative Assessment: teachers hearing what students know and adjusting the instruction accordingly

Importantly, the TRU framework situates student agency and voice alongside the coherence and the rigor of the math. Frameworks, including the TRU framework, related to CCSS-M or your own state standards can support you in your context to understand some of the core instructional shifts in balanced mathematics that are needed for success in detracked heterogeneous classrooms.

## LESSON LEARNED

Frameworks related to CCSS-M and other modernized state standards, including the TRU Framework, can support you in your context to understand some of the core instructional shifts in balanced mathematics that will support success in detracked heterogeneous classrooms.

Finally, in any curriculum you consider, you should look not only at the mathematics itself and the coherence of the curriculum but also the behaviors of mathematicians. The Standards for Mathematical Practice name these behaviors as the ways to make sense of and consolidate the content standards. The eight standards are the same from kindergarten through high school, so that all students are reasoning about mathematics in powerful ways. The standards for mathematical practice are as follows:

1. Make sense of problems and persevere in solving them

2. Reason abstractly and quantitatively

3. Construct viable arguments and critique the reasoning of others

4. Model with mathematics

5. Use appropriate tools strategically

6. Attend to precision

7. Look for and make use of structure

8. Look for and express regularity in repeated reasoning

(Common Core State Standard Initiative, 2021b)

Most states have adopted some form of these, and some states, such as Texas and Virginia, have their own version of process standards, which cover similar ideas though may be articulated a bit differently.

Any curriculum your school or district is utilizing, adapting, or adopting must provide multiple opportunities for students to be the thinkers and the doers of mathematics. Lesson structures that favor student voice and student sense making also need to position the teacher as the facilitator of learning and not the keeper of all mathematical ideas and answer-getting strategies. If the mathematical behaviors of students are not called out in the curriculum, this then becomes the work of teams of teachers as they collaborate.

## CONNECTING CURRICULUM TO TEACHERS IN DETRACKED MATH COURSES

SFUSD has always named its curriculum as one of the key levers for building a detracked math pathway with heterogeneous classrooms, alongside coaching, professional development, and the policy itself. The district names curriculum as a core lever because it holds that, as people, we are smarter together, that we learn more when we learn with others who think differently than we think, and that a powerful part of that learning is making sense of and building on each other's ways of thinking. To enact this belief, the students need math worth collaborating on, and teachers need to design lessons that provide for authentic peer collaboration.

When SFUSD leaders set out to reconsider its curriculum to align to the CCSS-M in 2012, there was not yet a set of instructional materials that was sufficiently aligned to the new content or practice standards. So in 2012, the SFUSD math department began creating a Math Core Curriculum, which foregrounded rich tasks and pedagogies that promote peer-to-peer discourse, as described earlier in this chapter. That is likely not your current context, and we do not want to make the case that you should start developing a curriculum. Your school or district likely has adopted or is considering adopting a curriculum written by a publisher that is used across classrooms and schools. Throughout this section, we share lessons learned that we believe could be helpful to other schools and districts working to maintain or adopt a curriculum and instructional materials that supports heterogeneous classrooms in a detracked math course system. We name the specific attributes that emerged for us throughout our own iterative cycles of development and revision, including having a curriculum that is

- task-based,

- prioritizes peer-to-peer interaction and student voice,

- values and empowers teacher leadership, and

- remains evolving and flexible to reflect new understanding about mathematics learning and the place of mathematics in the world.

These are hallmarks you may want to keep in mind as you move toward a curriculum that supports a detracked system of heterogeneous classrooms.

## Implementing a Task-Based Curriculum

In a task-based curriculum, each unit around a big mathematical idea features a series of tasks. Each task is an opportunity to examine the thinking of our students as their learning evolves: What did the students already know, what sense are students making of what they are learning, how are they applying what they have learned, and how are they consolidating that learning? Each task provides opportunities for formative assessment as teachers re-engage with the math throughout the unit. As outlined in Figure 5.1, the SFUSD Math Core Curriculum is one of many examples of a task-based curriculum.

### FIGURE 5.1   UNIT DESIGN OF THE SFUSD MATH CORE CURRICULUM

**SFUSD Core Curriculum Unit Design**

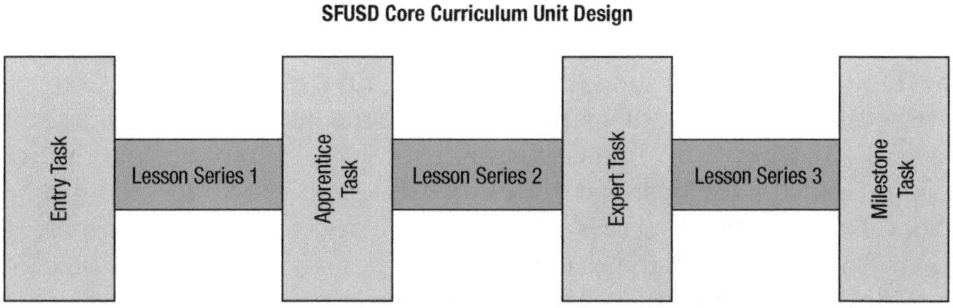

| Entry Task: | What do you already know? |
| Apprentice Task: | What sense are you making of what you are learning? |
| Expert Task: | How can you apply what you have learned so far to a new situation? |
| Milestone Task: | Did you learn what was expected of you from this unit? |

Source: SFUSD Math, adapted from Oakland Unified's Core Curriculum and Silicon Valley Mathematics Initiative

In the SFUSD context, the team as well as their colleagues across the Bay Bridge in the Oakland Unified School District also benefited from an ongoing partnership with the Strategic Education Research Partnership. This allowed them the opportunity to work with Harold Asturias, the director of the Center for Mathematics Excellence and Equity at the University of California, Berkeley, and Phil Daro, one of the lead authors of the CCSS-M. Asturias and Daro together supported the two district teams to better actualize the standards' commitment to conceptual understanding. They helped the district math leaders see the unit as the *right size* for developing key understandings, versus a single day's lesson, or an isolated experience such as a single math talk. Many Bay Area districts have also been member districts of the Silicon Valley Mathematics Initiative (SVMI), including a partnership with SVMI founder David Foster. These advisors were key in adopting this unit design shown in Figure 5.1, based on rich math tasks, which provides a structure for all the SFUSD Math Core Curriculum from prekindergarten through the Algebra 2 + Precalculus Compression course. Clearly, this was one context and set of circumstances; the key takeaway is the collaboration with partners.

## LESSON LEARNED

Consult with partners within your school or district such as other grade levels or departments, or outside of your school or district such as outside research partners or communities of educators-when making sense of or adapting key features in your curriculum.

## Prioritizing Peer-to-Peer Interaction

Along with the units, what pedagogical guidance does your curriculum offer? Are there structures that support students to interact in meaningful ways? In our context, to support the shifts required to promote the Common Core Standards for Mathematical Practice, the team developed a Math Teaching Toolkit (San Francisco Unified School District Mathematics Department, 2021b), inspired by our partnership with Oakland Unified School District, which serves as a resource for teachers to support an inquiry-based approach to learning mathematics with an emphasis on classroom discourse. While there are many components of the Math Teaching Toolkit, three signature strategies were selected and pulled forward to support classroom discourse specifically. These are:

1. **The Three-Read Protocol,** which is one way to do a close read of a complex math word problem or task and sets up students to ask questions

of the problem. This protocol aligns with commitments to literacy across the content areas and promotes access and rigor for language learners specifically (San Francisco Unified School District Mathematics Department, 2021d).

2. **Math talks,** which are a teacher-led, student-centered technique for building math thinking and discourse, for developing flexible thinking with number and shape, and for supporting students to listen to and build on the thinking of others (San Francisco Unified School District Mathematics Department, 2021a).

3. **Participation Quiz (secondary) or Groupwork Feedback (elementary),** which are strategies to help establish or reinforce norms for groupwork in a cooperative environment (San Francisco Unified School District Mathematics Department, 2021c).

Throughout the units the team developed, they called out opportunities to leverage these and other strategies, offering many examples from their own practitioners.

## LESSONS LEARNED

Your curriculum must foreground student voice if the students are to enact the behaviors of mathematicians as described by your state's mathematical practice or process standards.

### Valuing Teacher Voice

*Working with a team of teachers to write math curricula this week has been an amazing experience that has . . . opened my eyes to the potential and true intent of the Common Core. It's not about constant test taking [or] seemingly impossible homework questions developed by people trying to fit new standards into an old format. It's about developing deep understanding, respectfully challenging ourselves and one another, and not being afraid to tackle difficult . . . problems with innovative approaches. These are the skills that our children need [to] learn to be resilient and responsibly interdependent in this ever more complex and fragile world.*

*—Lauren, SFUSD kindergarten teacher*

In SFUSD, educators revised their curriculum over many years based on feedback from the field, growing toward greater horizontal and vertical alignment, building out multilingual resources and technology recommendations. Over many years, SFUSD has revised curricular resources including individual tasks or lessons to support language goals, to provide for more groupworthiness, to move away from language or contexts that may limit multiple means of representation, and to rethink context bias. This was our work; in your context, you may already have a rich and relevant curriculum that is standards aligned, or you may be adopting curriculum from across a range of high-quality materials that are now available to math educators. The key takeaway is, when considering curricular choices, you must listen to and work with teachers to ensure your curriculum continues to be responsive to both mathematical understanding and also to enacting your detracking policy.

## LESSON LEARNED

As you create, adopt, or adapt your school or district math curriculum, ensure teachers are leading voices in this work.

## Learning About and Improving Math Curriculum as Contexts Change

As educators, we all strive toward math achievement for each and every student. Before the global COVID-19 pandemic and calls for racial justice after the murder of George Floyd, SFUSD and districts across California and across the world were already grappling with questions of humanizing and culturally sustaining pedagogy and interrupting structural racism in mathematics. Culturally sustaining pedagogy lifts joy and beauty and celebrates people of all backgrounds (Paris, 2012). In San Francisco and many districts, district leadership has charged all educators with learning about and implementing humanizing and anti-racist practices, to bring into coaching conversations and professional learning spaces. This can allow opportunities for schools, teachers, and students to use mathematics to recognize and solve social inequities that occur in our communities. This practice is consistent with SFUSD's core values (San Francisco Unified School District, 2021) and with the CCSS Math Standards themselves, such as Practice 4 Modeling (Common Core State Standards Initiatives, 2021b), which states, "Mathematically proficient students can apply the mathematics they know to solve problems arising in everyday life, society, and the workplace." One curricular resource that attends directly to social justice in mathematics is

*High School Mathematics Lessons to Explore, Understand, and Respond to Social Injustice* (Berry et al., 2020).

Many organizations have explicitly called us as educators to name and dismantle structural racism in mathematics, and in education more broadly. These include, but are not limited to:

- The Benjamin Banneker Association Social Justice (2017) position statement
- NCSM and TODOS (2016) Position Social Justice position statement
- California Mathematics Council (2021) position statement
- *A Pathway to Equitable Math Instruction: Dismantling Racism in Mathematics* (May 2021), which is a workbook designed to support educators to grapple with these ideas together (https://equitablemath.org/wp-content/uploads/sites/2/2020/11/1_STRIDE1.pdf)

Achievement for all requires a social justice stance.

Before the COVID-19 pandemic, communities of math educators across the United States were taking up the call to look at our math classrooms, at the lived experiences of our students, and at the systems and structures that support learners to find success or to experience harm. Through the pandemic and now that we are either in schools in person or adopting to hybrid models of instruction, it has become even more clear that there is humanizing work to be done, where relationships matter, and we attend to each other's health and well-being alongside our commitment to developing the mathematical brilliance of each and every learner. The team in SFUSD is still engaging in this work of more authentically thinking through revisions to its curriculum that will support educators and students to enact humanizing pedagogy and anti-racist practices. They expect this work to continue for many years to come.

## Questions to Consider for Your Context

▶ As you've learned in this chapter, your curriculum plays a role in the instruction experienced within each detracked math classroom. The curriculum will either help or hinder teachers as they work with more heterogeneous grouping. Here we provide some practical questions you and your team can explore to help you think about your curriculum and what (if any) work you need to do to prepare your curriculum as you detrack your math courses.

• What is the current curriculum you are using? What are the curriculum's key features? How will those features relate to students in heterogeneous classrooms with a mixture of backgrounds and strengths?

• How does your curriculum support heterogeneous math classrooms? How many points of access do the math tasks and activities designed provide for students?

• How do teachers think about and use your curriculum? How do students engage with the curriculum? Does the curriculum help teachers and students engage in more task-based, collaborative math activities? If not, are there ways your curriculum could be adapted or supplemented to support for differentiation and collaboration?

## Activity 5: A Collaborative Read of a Progressions Document

The Common Core State Standards in mathematics were built on progressions: narrative documents describing the progression of a topic across a number of grade levels, informed both by research on children's cognitive development and by the logical structure of mathematics. These documents were spliced together and then sliced into grade level standards. From that point on, the work focused on refining and revising the grade level standards.

http://ime.math.arizona.edu/progressions/

▶ For this activity, you will engage in a collaborative read of one of the Progressions Documents; these documents could support alignment work with how important mathematical ideas build even if your state has not adopted the Common Core State Standards. NCTM also has a series about essential understandings that could similarly support your learning within any domain across grade levels (https://www.nctm.org/store/eu/).

In our SFUSD context, the Progressions were a valuable resource not only for curriculum development with our teacher developers and within our own team but also for ongoing professional learning districtwide and within individual school buildings. This activity allows you to make sense of the big ideas in the standards and also consider next steps in your curriculum journey. This activity can also support teachers to understand the different ways that math concepts and skills are connected so that they can use students' prior understandings to build from versus remediating with standards from younger grade levels.

You can select from across the range of documents at http://ime.math.arizona.edu/progressions/ based on a variety of factors: A grade-level team might want to more deeply understand what came before or what comes after their own content standards to support their planning and instructional choices, or a group of site leaders or district leaders who may not specialize in math might want to capture for themselves a bird's eye question such as "What are the big ideas in middle school math?" These documents tell a narrative of how one domain will build across grades, offering grade-level examples and visual models throughout.

Once you have selected the progression you will read collaboratively, choose a reading protocol that will keep your team on track. These documents range in length from about 12 to about 30 pages, with most 15 to 20 pages long; the length of the progression you choose might determine the protocol you choose. No matter how you agree to read, these are good questions to have in mind:

- What thread or patterns do you see across the grade levels? Where can you find evidence of coherence as ideas build across the grades?

- What are the implications for your curriculum? What will you adopt, adapt, or abandon based on your reading?

- What are the implications for your classroom, your school, or your district, beyond curriculum? Might these documents support different learning models or support you to refine a vision of mathematics teaching and learning?

If you as a community already have reading protocols you leverage, by all means start with those familiar practices. If you do not yet have a practice of collaborative reading, or want to try something new, here are some reading protocols you can choose from:

- **Notice and Wonder:** Each reader can choose something they notice and something they wonder, a practice that closely mirrors math talks teachers may already be using with students (https://www.nctm.org/noticeandwonder/). Each reader then shares their quotes in a round robin using the sentence stems "One thing I notice . . ." and "One thing I wonder. . . ." After each reader has shared, the group can decide where to go more deeply, building on the thinking of peers and asking questions until ideas make sense. While this may or may not lead you to action steps, it is a great way to make sense of new text together.

*(Continued)*

*(Continued)*

- **Progressions trace by grade:** Each reader can describe in depth a big idea from a grade level, for example on the 3rd–5th Progression on Number and Operations—Fractions, one reader can describe or demonstrate a 3rd-grade idea, one reader can describe or demonstrate the 4th-grade idea that builds, and one the 5th-grade idea that builds further. Variations:

  - A progressions trace could also lend itself to a jigsaw where different group members read different parts of any given text then share back.

  - Each team member could actively look into your curriculum for how these big ideas build within the instructional materials you are using or considering.

- **Punctuation notation:** First, each reader will read the progression individually, noting throughout by marking up the text with punctuation marks:

  - An explanation point for something you are enthusiastic about or perhaps hadn't thought about in this way

  - A comma where you want to know more or explore further

  - A question mark where you might disagree or want to argue back with the text

  - A period when you come to an idea that you already feel comfortable with or have already thought about in this way

  Then when you gather as a team, you can share virtually on a platform such as Jamboard, or physically on posters, some of the punctuation points you found individually. You are likely to see patterns as a group, as well as find interesting points of discussion if team members used different notation for the same idea. Where might your discussion take you if one team member is excited by something that another team member disagrees with? There are many different versions of this reading protocol, which your school or district may already be using for literacy instruction.

The goal of any shared reading is to grow your professional efficacy as a team, as well as your clarity on the coherence of the mathematics. The Common Core Progressions Documents are a great resource in that work.

# CHAPTER 6

........................

# PROFESSIONAL DEVELOPMENT TO SUPPORT HETEROGENEOUS CLASSROOMS

The district mathematics team has gathered teachers from the various sites for professional development day, as they do a few times a year. Today, Stephanie Hanson was attending the Algebra 1 Math Collaboration Day professional development with Algebra 1 teachers from across the district. Stephanie enjoyed these days, because she and her colleagues were able to put the hustle and bustle of schooling aside and spend time focusing on diving deep into a few math lessons. She walked in, said hello to colleagues she has gotten to know over the years, and from the directions being projected, she found a table to sit with three other people from different school sites. As a district learning community, they would do math from upcoming units, reflect on the experiences, and plan units and lessons with their school team.

After the district facilitators began the day, it was time to do some math. The district leaders facilitating the day asked the teachers to put on their student hats. Stephanie was happy to jump in and put her teacher hat aside. On this day, Stephanie and the other teachers attending were working on math tasks from the spring units on quadratic functions. The district facilitator wanted the teachers to see the big ideas of the mathematics across multiple units. The focus of this day, as always, was to support teacher collaboration, to understand a projection of learning over time, and experience the mathematics from a student perspective. This would then support them to create learning experiences for students to make sense of problems, make connections, and justify their thinking in a collaborative way.

The facilitator launched the math task by sharing the doing math norms (Figure 6.1) and exploring some of the ways the students could be smart in this task, prepping participants to think about what strengths they can bring to their group conversations today.

FIGURE 6.1  DOING MATH NORMS

## Doing Math Norms

- Stay together as a group.

- Work in middle space.

- Share your thinking out loud.

- Use strategies your students might use (consider prior knowledge, different approaches).

*For this lesson, each group of students was given cards for a math task with different pieces of information on separate cards. The directions from the district facilitator said: As a group, choose one card at a time to discuss. Work together to complete the graph and/or table for each situation. Then consider the questions together about this situation. Stephanie's group got to work, sticking to one conversation as a group and making deep connections along the way.*

*After the math task, the participants were given time to debrief the experience with their teacher hats. This year, the district math leadership team had chosen a theme of Access and Rigor. Figure 6.2 shows the Access and Rigor definition developed by the central math team as well as the debrief questions teachers were asked to consider.*

FIGURE 6.2  ACCESS AND RIGOR DEFINITION & DEBRIEF QUESTIONS

**Access** means all students have a way to participate in the learning experience and use their strengths to be successful in the grade-level math in the task.

**Rigor** means all students are challenged and are deeply learning the grade-level math objectives and big ideas.

**Keeping equity at the center** means considering access and rigor together for all students, including how to build off of and foster students' strengths.

SFUSD Math

Debrief Questions:

1. What were the big ideas of this lesson?

2. How did the task design (cutting up cards, colored pencils, etc.) support **access** to the mathematics?

3. How did the task design and teacher moves support **rigor** in the mathematics?

*During the teacher hat debrief, the district facilitator started by asking someone to read the definition of Access and Rigor aloud, then the teachers discussed the discussion questions in their small group, followed by groups sharing out. In Stephanie's group, they engaged in a deep conversation, building off what the others noticed in the math lesson. Some of her colleagues noticed how the cards focused the group and allowed each person to participate. Others noticed how the facilitator's teacher moves and lesson structures held the groups accountable to the doing math norms and holding their group to one group conversation. Stephanie noticed that while the facilitator walked around listening to groups during the math, she used questions that pushed for justifications. She also noticed the facilitator referring people back to their group when they had questions. At the end of the teacher hat discussion, Stephanie remembered the district facilitator reemphasizing: "So access and rigor does not mean access for some and rigor for others. Instead, there were a variety of ways that allowed access for all students to the task and all students were able to grapple and make sense of rigorous mathematics."*

*After lunch, Stephanie was happy to dive into some deep planning with the other Algebra 1 teachers at her site. They started by checking in with one another and making a plan as a team for which lessons to look at. Someone brought up how good it was to remember structures like putting information on cards and giving students strict roles that can support more participation. Someone else acknowledged how these days are good reminders to do math together as a team to help them decide the lesson structures to add. The team made a plan for two upcoming lessons to do math together, and they got to work, grabbing some scratch paper and having someone read the task out loud to get started.*

## WHAT TO EXPECT IN THIS CHAPTER

*Having a detracked, heterogeneous mathematics pathway requires supporting teachers in the creation of equitable classrooms so all students have access to learning rigorous grade-level mathematics. In this chapter you will*

- *Explore how important it is to explicitly attend to vision building in professional development and ways in which you might consider supporting educators to visualize these visions, expanding what is valued in mathematics, expanding our ideas of what students can do mathematically, and considering more pedagogical strategies that support equitable student voice in classes.*

- *Recognize how to design professional learning opportunities that support the building of sustainable site learning communities to support teachers to continue to reflect on the work with their site colleagues, including interrogating their own belief systems, while working to provide all students with rich learning experiences.*

- *Consider how professional learning opportunities can support teachers by connecting to other district initiatives, professional developments, and the strengths within the community.*

*This chapter uses the terms* professional development *and* professional learning opportunities *interchangeably. These terms are used to mean bringing a group of educators together for learning. We believe you can apply these big ideas to any group of educators you bring together, including a grade- or course-level team, administrators, teachers from across the district, or even your central leadership mathematics team.*

## QUESTIONS TO CONSIDER WHILE READING THIS CHAPTER

All professional developments provide experiences that are either explicitly or implicitly conveying a particular vision, about what is important related to mathematics and mathematics teaching and learning. Here are some essential questions that are important to consider as you provide professional development experiences to support educators.

- **Vision of a Mathematics Classroom:** What is the vision for an equity-based mathematics classroom that you are hoping your students will experience in your mathematics program?

- **Vision of the Adult Learning Community:** How is your professional development supporting the vision of a powerful adult learning community that you want to see at your site(s)?

- **Working From Strengths:** How is your professional development recognizing and building on the strengths of your students and teachers?

- **Teacher Reflection:** How is your professional development making space for teacher reflection and introspection related to the equity-based vision you are working toward?

- **Making Connections:** How are your professional learning opportunities intentionally making connections for teachers?

## BIG IDEAS OF DESIGNING PROFESSIONAL DEVELOPMENT

In Stephanie's story at the beginning of this chapter, there is a strong learning community of educators that was created with teachers across the San Francisco Unified School District (SFUSD). Teachers felt excited to attend and were often reinvigorated after leaving a professional development. When speaking to teachers throughout the district, you can hear the similarities about how they talk about students: truly believing that all their students are mathematically brilliant, seeing the value in collaboration amongst adults, knowing that doing math together supports them to be more prepared to see all students' strengths, and grappling with ideas for how to support all students to be successful in mathematics courses. One of the reasons that San Francisco has built a robust community where there is a continuation of the strong beliefs in our students' mathematical abilities, attention to equity, and the similarities in practices across classrooms is due to, we believe, the professional learning opportunities that the central mathematics team created for teachers across a very large district. The following gives examples of how SFUSD has used professional development to build these visions, some big ideas of strengths and reflections as a part of the professional development designs, and the intentional connections among all professional learning opportunities.

### Centering on a Vision for Equity-Based Mathematics Classrooms

Professional development models, at their heart, move teachers and practice toward a vision. Those running professional development opportunities or looking to bring professional development into their system need to consider the vision they are trying to move educators toward. *What is the vision for an equity-based mathematics classroom that you are hoping your students will experience in your mathematics program?* One response might be, "We want a discourse-rich environment for our students." Are all your facilitators clear about what this vision looks and sounds like? How can your team become more coherent on this vision?

In San Francisco, the central mathematics team has a district vision statement for the mathematics classroom: *All students will make sense of rigorous mathematics in ways that are creative, interactive, and relevant in heterogeneous classrooms.* Over time, the team added two additional premises to share along with the vision statement to help participants understand the underlying beliefs that informed design choices and facilitation moves. The premises are: *All students are mathematically brilliant* and *math is a web (not a ladder)*, meaning that there are many ways to do and understand mathematics (Featherstone et al., 2011). The vision statement and these premises—both of which were introduced in Chapter 3—explicitly highlight important components that need to be attended to in professional development: beliefs about students, about mathematics, and about how students interact with the mathematics. However, a vision statement is a beginning and is not enough to change teacher beliefs or practices about teaching and learning mathematics. The facilitation team needs to provide experiences to help teachers internalize what the vision looks and sounds like. This can happen in multiple ways. Here are three learning experiences the SFUSD team implemented.

Having educators **do math together,** facilitated by someone who can lead the experience with pedagogical strategies you would hope teachers would lead with students and that can help us view math as expansive.

Having educators **watch a video of students** in a mathematics classroom that shows the vision of what we are looking for in our mathematics classrooms.

**Observing classrooms in action.** Looking for the many strengths in the classroom, based on evidence from students' ways of thinking.

## Doing Math Together

In the vignette at the beginning of this chapter, Stephanie and her colleagues were experiencing what it is like to be a learner in a discourse-rich mathematics classroom through doing mathematics together with their student hats. This experience allows teachers to see math as expansive, as a web. On this day, teachers learned to experience a task that asked students to make connections among multiple representations, pushed for justifications, and showed how there were many ways to approach the task. When asked by facilitators for feedback about these professional developments, Stephanie said, "The best part about those [professional developments] was always learning from you in the way you handled us even when things were not explicitly said. You modeled things, the participation structures you used and the groupings." A middle school teacher, Michelle, also referred to the experiences of doing the math along with the time they were given after to debrief the mathematics experience in their teacher hats. Michelle said, "I appreciate the fact that we have to experience this as learners and then y'all push our thinking to really think about what kind of impact will this have on our babies. That's important." Participating in a class-like setting where the facilitator takes on the teacher role using rich mathematical tasks along with pedagogical strategies that bring more student voice into the room and more equitable participation creates a vision that teachers will reflect on. This then allows teachers to think of ways to create their own space that showcases the many mathematical brilliances each student brings.

## Watching Video of Students

Watching classroom video can be another powerful learning tool to experience the vision of a mathematics classroom. The SFUSD Complex Instruction program includes the usage of video clubs, using a model developed by Jilk and O'Connell (2014). The goal of this Video Club is to support a community of teachers to build a common vision of teaching and learning based on encouraging equitable participation in small group work (Jilk & O'Connell, 2014, p. 223). The videos chosen help participants see the vision of what is possible for *who* can be smart at mathematics, *what* math can look and sound like, and *how* we want students to participate with each other and with the math classes. First, to push and reinforce the belief that all students are mathematically brilliant, consider actively capturing video of students who have traditionally been underserved in your context, students of color, neurodiverse students, multilingual students, and so on. Second, use video that features a math task that is rich and rigorous, expanding the idea of what it means to do mathematics, being more than just getting right answers quickly and possibly requiring multiple ways of thinking, encouraging justification, asking questions, and using manipulatives. Third, capture video that highlights student voices in structures that match your vision. For the San Francisco team, they made sure to use video focusing on one or two groups of three to four students engaging with mathematics and each other with equitable participation, based on the vision they are working with teachers to achieve.

Next to capturing the right video is learning how to frame and facilitate the video watching in the professional learning opportunities. The protocol from Jilk & O'Connell focuses teachers' noticing on students' strengths rather than deficits (Jilk, 2016; Jilk & O'Connell, 2014). Our world makes it easy to pick out what one deems needs to be fixed. It takes strong community agreements and facilitation of strict protocols to keep participants to the strengths they notice. In the San Francisco example, teachers state what they notice around student understandings and participation norms that supported learning in the video. Hearing the many pieces of evidence and the variety of language participants use to describe students' strengths broadens what teachers notice and value in their classroom, leading to students "shift[ing] their own beliefs about their strengths as mathematical learners" (Jilk, 2016, p. 189). A key takeaway from running Video Clubs is how the use of video and the facilitation focused on strengths rather than deficits can be brought into a variety of professional development contexts to build a coherent vision of mathematics teaching and learning.

## Observing Classrooms in Action

Observing classrooms in action is another way to help teachers get clearer about the vision of an equity-based mathematics classroom. Professional learning opportunities that include classroom observations in the learning are those such as lesson studies, peer reciprocal observations, or learning walks. In all these professional learning opportunities, educator teams set themselves in the position of learners, preparing to learn about student thinking. Often, teacher teams develop a problem of practice to investigate and use the evidence they capture to learn more about their question. All these structures have participants taking the time to do the math they will soon see students doing with colleagues prior to the observation. This process prepares observers to notice evidence of how students approach the mathematics, their ways of thinking, and the many different strengths students bring to the conversations. It is important that as colleagues are opening their classrooms and practice for others to see, facilitators and leaders focus the team to look for strengths. In addition to creating a safe and brave learning environment, focusing on strengths allows us to find the pockets of where our vision is happening for the team to learn from. The debrief of the observation allows for participants to share the evidence and consider next steps to build from these strengths in the future for all their classrooms.

Successful heterogeneous classes have the space for all students to participate equitably, where every student learns rich grade-level content at a deep level. It is essential to support teachers to work toward equitable participation in classes and to become more aware of the status characteristics, such as race or ethnicity, gender, age, height, courses a student is taking, and so on, that in your context have often led students and educators to see particular people as smart and capable. According to Cohen and Lotan (2014), "a status characteristic is an agreed-upon social ranking where everyone feels it is better to have a high rank than a low rank" (p. 34). Observations can give the educator team a lens to focus on the strengths of students, especially focal students, who belong to groups that have been underserved by your systems. Attending to these status characteristics as you gather evidence in classrooms can support a team of educators to determine curriculum choices, teacher moves, and instructional routines that support students to participate. Working toward this vision requires that professional learning supports this vision in all capacities and that everyone is aware of the societal status characteristics that, if left unaddressed, cause a repeat of the status quo in which people who succeed belong to the groups who find privilege in our society.

> *Successful heterogeneous classes have the space for all students to participate equitably, where every student learns rich grade-level content at a deep level.*

## LESSON LEARNED

Keep equity at the center. Stay clear and upfront about working to support all students as you are building equitable classrooms. As you center your vision of what is possible for students, be sure to give examples, especially showcasing and highlighting the brilliance of Black students, students of color, and other marginalized groups.

## Centering a Vision for Adult Learning Communities

Creating equitable classrooms for all students is difficult work and requires rich problem solving and collaboration among colleagues. Professional learning opportunities can be built to support collaboration, intentionally considering the adult learning community as the unit of change. *How is your professional development supporting the vision of a powerful adult learning community that you want to see at your site(s)?*

Think of the sites where teachers enjoy the community and stay for many years in their career at one site—sites where you see similarities in classrooms where students of all backgrounds are excited by mathematics learning and see similar instructional moves by teachers. These similarities do not happen through isolation. Rather, there is strong collaboration among the teaching team to support a similar vision. Teachers and teacher leaders are committed to collaboration and learning through ideas such as doing mathematics together, analyzing data and student work together, reflecting on evidence from classroom observations or video, and noticing and reflecting with colleagues. They have structures set up to learn together such as common planning time, professional learning time, peer-reciprocal observations, lesson study, video clubs, or even a department- or grade-level meeting. They consider important systemic setups such as apprenticing in new teachers, distributive leadership, and how the school schedule can support planning times. Additionally, these teams are always searching for better practices to support equity, looking for ways to support all students across the math program.

Those providing professional learning opportunities must intentionally create experiences that center the vision of a strong learning community, give teachers and leaders ideas and tools for what this learning can look like, and demonstrate how teams can re-create these experiences at their sites. To support the building and

> *To support the building and sustaining of strong site learning communities, professional learning opportunities must position the learning community as the unit of change to truly create the desired change in your site or district.*

sustaining of strong site learning communities, professional learning opportunities must position the learning community as the unit of change to truly create the desired change in your site or district. Here we consider three types of professional learning opportunities, working with different groups of teachers:

- Working with educators across sites or in **districtwide** professional developments

- Working with **teacher leaders** in professional developments

- Supporting **individual teachers**

This section will end discussing the importance of attending to **equitable participation** in professional development and in teacher teams.

## *Districtwide Professional Development*

As you read the vignette and the ideas around vision building through doing mathematics with colleagues, watching video, and observing classes, you can hear the intention to create collaborative cultures both in a larger network and within site teams. Support providers are not able to be in a class every day, let alone at a site daily, and there is often a sunsetting date for support due to funding sources. Thus, professional development should always be working toward giving the teachers the autonomy and structures where they can build a strong learning community that can be sustained over time, without support.

The Collaboration Days used in San Francisco, and inherent in the title of the day, reminds educators and administrators that collaboration is at the heart of the work. These days were originally modeled from professional development in the Complex Instruction program supporting both site and cross-site collaboration (Jilk & O'Connell, 2014). The central math team had secondary mathematics teachers attend with their grade- or course-level teams in a cross-site day of learning where they were immersed in activities that allowed them to experience the vision of a strong learning community, both for students and adults. One way the San Francisco team gave teachers the experience of the vision of a strong collaborative team was through unit and planning protocols. One of these, a Lesson Planning Process, guides a team to create a fully thought-out lesson. You can see this Lesson Planning Process in the Activity at the end of this chapter, and we hope you use this with a team to plan a future lesson. During the teacher hat debrief of the mathematics task in a Collaboration Day, the facilitators were able to describe moving through the Lesson Planning Process to see the result that the participants experienced. This gave a connection to the planning time in the second half of the day where the teams were asked to use the Lesson Planning Process with upcoming lessons in the curriculum. Teachers have found this process useful and have brought it back to their site teams to use during their planning time.

## LESSON LEARNED

To create sustainable change, it is essential to consider the site learning community as the unit of change and build professional development with this in mind.

### Teacher Leadership

Additional structures in San Francisco at the secondary level were ongoing teacher leader professional development. These were often small teams of teacher leaders per site, sometimes with department leads, course leads, or as teacher leaders in a particular program, all coming together with other sites. Teacher leaders were asked to come in pairs or trios, so they had a partner to think through the ideas to bring to their site. Inherent to all these structures was the idea of supporting teacher leaders to be facilitators of learning back at their sites, moving beyond leading in a logistical manner. The facilitative team designed professional learning for leaders to see the power of doing math together, discuss issues related to equity, consider important structural ideas like school schedules, have teams create goals, and create activities that teacher leaders could then bring back to their site teams. Stephanie from our vignette became a department chair and has told the initial facilitators, "We copied how you ran the Department Chair meetings [with our department]. We would do math together, try to have something that was new and equity focused, like a learning experience together, and then go into logistics."

The central team knew that leaders need to see and experience the vision of a strong adult learning community. In a meeting with one group of teacher leaders, called *teacher facilitators*, the central team brought a reading to help them visualize a particular mathematics department that had created an example of the vision. On this occasion, the teacher facilitators were asked to use the reading from Cabana et al. (2014) to develop a vision for departmental learning together. You can see in Figure 6.3 the Vision of an Adult Learning Community that our teacher facilitators came up with.

FIGURE 6.3   VISION OF AN ADULT LEARNING COMMUNITY

All teachers in a learning community will work together to

- Continue to deepen our practice of collaboration

- Treat teaching as groupworthy

- Take collective responsibility for the department and all students, holding high expectations and a common vision for what it means to be a math learner

- Continue to work toward strengthening collective knowledge of content and of teaching within and across grade levels through reflection and refinement

- Each year, attend to community safety (to be vulnerable) and to build on each other's unique strengths so that our team grows and develops, working toward sustainability

Source: Developed in 2016 by the SFUSD Complex Instruction Teacher Facilitator community after reading Cabana, C., Shreve, B., & Woodbury, E. (2014).

The central mathematics team in San Francisco has then continued to use this vision statement with other groups of leaders, including department chairs and administrators, to center their work.

## LESSON LEARNED

Similar to supporting teachers to see the vision in the mathematics classroom, it is important to help teachers and teacher leaders visualize and experience the vision of an adult learning community. It helps if you (1) start by naming the vision statement, (2) give the teachers and teacher leaders an experience of the visions, and (3) make space to debrief, reflect on the experience, and consider how to take it into practice.

*Working With Individuals*

Keeping the learning communities in mind any time we are working with teachers can be difficult, especially when we are thinking about supporting individual teachers through professional development or even one-on-one times such as coaching. Ways to support the learning community in a coaching relationship might sound like:

- "How are you and your grade-level or course team(s) collaborating?"

- "Have you asked your colleagues about what they do for ___?"

- "(Teacher on their team) is really good at ____; they might be someone to think through ideas with."

- "Have you observed ___? You should see if they would be open to you sitting in the back during your prep to get ideas."

- "You and your colleagues should attend the local conference together. Attending together and sharing what you learn is a wonderful team bonding and learning experience. Check in with your department chair or principal to see if a team can go."

Just like with students, the language we use with teachers can encourage and set the stage for how teachers plan together. Even in speaking with an administrator, one can speak to this vision, that we need strong learning communities to move the work forward, or our words can support individualism and siloed teachers. For example, "Principal_____, do you know if your Math 7 team is signed up to join this professional development day together?" versus "Please send this link to your teachers to sign up" is one small way to update language to assume teachers are collaborating and participating in learning events as a team.

## LESSON LEARNED

As you work for sustainable change, it is essential that anyone supporting teachers works to see how their support and work with individuals will impact and connect to the work of the larger site learning community.

Similar to our work with students, we need to be aware of how status issues can seep into any group working together, interrupting equitable participation, and not allowing the group to reach the potential of learning from all on the team. As the San Francisco team worked with partners who deeply understood, researched, and practiced complex instruction and who therefore attended to equitable participation and status in any group, the central math team learned about what to attend to when working with adults and ways to consider how to mitigate for the status in the room and potential issues that might arise. With adults, there are many characteristics that might impact how people view each other and thus how they communicate and work together. This can show up with one person being the one everyone goes to for help, someone who often dominates conversations, someone whose ideas get overlooked, and so on. Decisions about how to collaborate and who is competent or who is not competent often come from *perceptions* of competency. Similar to students, race, culture, gender, gender expression, sexual orientation, primary language, and other characteristics play a role in these perceptions. With teachers, you have a larger age span, so sometimes the eldest are seen as most competent and other times it is the younger generation that is seen as more competent. You also have the courses that one teaches (i.e., are Calculus teachers seen as *smarter* than 9th-grade teachers?) or the degrees and credentials one has (i.e., some states have foundational credentials to teach through 10th grade, while others have degrees in mathematics). You also have longevity and additional leadership roles in a school (i.e., in one school the athletic programs have a lot of status so a mathematics teacher who also coaches a sport might be seen as the main one with good ideas). If one member feels silenced or undervalued, the community is broken and the team loses out on their brilliance.

All these qualities play out through participation, so as facilitators of professional learning, it is important that we create structures that allow all teachers to be seen as competent and all to have an opportunity to speak and be heard. While norms and agreements are helpful to set the stage for equitable participation, it is also essential to add participation structures to ensure that participation is equal. The San Francisco team often uses roles with teachers, like those teachers might use with small student groups (i.e., Reporter Recorder, Team Captain, Resource Manager, Facilitator). This goes beyond the mathematics experience to teacher hat times where a debrief will ask a small group to begin the conversation having each person share out in a round robin fashion, starting with a particular role, for example, the Team Captain. As the facilitators are watching conversations, they are looking for participation similar to students, noticing when there is more equitable participation and when the participation is not equal. They will walk around and notice if something, like a water bottle, is in the center of the table, blocking eyesight to what they are discussing, and ask for the bottle to be moved to the edge of the table. They might ask for the Recorder Reporters to share out group ideas from each group so that they represent all members' ideas and it isn't always the same voices. They might notice that in one group, the Resource Manager, an older teacher whose primary language is not English, has not spoken much in their group, so the facilitators ask that in the next round for all Resource Managers to start their group conversation. The facilitation team hopes that this modeling and attention to equitable participation will encourage teachers and leaders to put more participation structures in place for their team's learning time and ultimately support their understanding that can provide similar equitable learning experiences for their students.

## LESSON LEARNED

Working to provide equitable participation experiences with adults during professional development is as important as providing equitable participation experiences for our students, to learn from all the strengths within your community.

## Recognizing and Building on the Strengths of Your Students and Teachers

An equity-focused facilitation team sees a community in the classroom and in a community of teachers as a collective entity, where everyone brings expertise and everyone has something to learn. As you have read this book, we hope you picked up on the strengths-based language and attention to both students and teachers. Many researchers and experts in the field of mathematics education are helping educators refocus their attention to strengths (Jilk, 2016; Skinner et al., 2019; White et al., 2018) and move away from the deficit thinking that we have all been indoctrinated into that comes from negative stereotypes about students of color and other marginalized groups (Adiredja & Louie, 2020; Seda & Brown, 2021, pp. 66–70). By viewing our students and teachers as having strengths to offer, we are modeling the creation of a community where no one person knows it all and where we are collectively smarter when we make space to see and learn from each of our strengths. *How is your professional development recognizing and building on the strengths of your students and teachers?*

> *An equity-focused facilitation team sees a community in the classroom and in a community of teachers as a collective entity, where everyone brings expertise and everyone has something to learn.*

Let's return to the premises of the San Francisco mathematics team: *All students are mathematically brilliant*, and *math is a web (not a ladder)*. This first premise articulates the idea that our students bring mathematical strengths to our classrooms. When the central mathematics team decided on the first premise, they intentionally used the word *brilliant* as an aspirational term. It is simple enough to say that all students are capable of learning mathematics and make excuses for those students who don't seem to want to; but to aspire to see a student's brilliance is to say, "I know this student has mathematical strengths, I just need to give them opportunities to shine. I need to go on that treasure hunt to find their strengths!" We must believe this for all students, otherwise it is too easy for biases to play out.

The second premise, *math is a web (not a ladder)*, refers to the idea that doing mathematics is expansive. This means that there are many ways to approach, interpret, and make sense of mathematics. This also reminds us that mathematical topics are interconnected in ways that are nonlinear, that we can learn at a deeper level of understanding when we make these connections. For example, if a student forgets a previously taught topic, there are other topics, experiences, and knowledge we can build from to help them dive into the content. When mathematics is taught only in a linear way, where students cannot move on until they have mastered a concept or skill, then only some students are able to experience success. Seeing mathematics as expansive, as a web, allows more opportunities for our students to all be smart and brilliant. It is impossible to see all our students as mathematically brilliant without seeing math as a web.

Those leading and designing professional learning opportunities must attend to these premises to help educators, the community, and administrators unpack the biases and stereotypes that lead to deficit thinking of students. It is important to intentionally design experiences for teachers to see that our Black and brown students, students whose primary home language is not English, and students who are neurodiverse are brilliant at the challenging, grade-level mathematics. Strong facilitation with the use of premises and protocols can support teachers to be prepared to notice the many strengths students bring and how we can use these ways of thinking to provide deeper connections among students' understanding.

Similarly, we must observe our colleagues with a strengths-based lens. Additional premises that the San Francisco team has about teachers: *All teachers have strengths in teaching and learning*, and *all teachers care about their students and their practice*. These premises ground the facilitation team in creating professional learning opportunities knowing and recognizing the strengths and expertise in the room. Language such as "These teachers need . . ." can position those facilitating professional development as the ones with all the expertise and power in the room. Contrary to this, professional developers should recognize that there are many strengths among those in the room, including the educators in the room.

The pedagogy of complex instruction has a very powerful strategy of *assigning competence*, publicly naming the intellectual strengths of students and why this strength is important to the group's learning, to disrupt the perceptions that students fall into a hierarchy of who is *smartest* (Cohen & Lotan, 2014, pp. 156–160; Featherstone et al., 2011, pp. 87–92). We can apply this strategy as we work with teachers. Authentically publicly naming strengths of teachers in front of other colleagues supports teachers to feel more confident in themselves, to recognize the strengths in colleagues, and to experience what is valued in your vision of teaching, learning, and collaboration. Assigning competence also helps build relationships in the community, between teachers, and between facilitators of professional development and teachers. Building strong relationships creates a safe and brave space for teacher reflection and learning. Many of us have had moments when a colleague named a way they noticed how you benefit the team. The more specific they name it, the more you learn something new about yourself that you maybe had not yet realized or articulated in that way. And the power of hearing a strength named in front of colleagues heightens your confidence and the way someone else sees your value. Consider how often you look for your colleagues' strengths? Can you name how each colleague is a strong teacher or colleague?

In our professional developments, the San Francisco team has tasked teachers and teacher leaders to list students and/or colleagues and write down the strengths you know about them. Notice areas you know specifics of someone's strengths. Notice areas where you only know general strengths. Notice where it is difficult to name strengths, and consider the reasons why. Then we ask educators to make it a goal to seek opportunities to search for strengths (i.e., sit with a colleague you don't normally sit with to do math, conduct an interview with a focal student, create a task that brings in a student's geometric strength).

## LESSON LEARNED

Attending to strengths, from students to educators to families, is essential in creating change so you can build from the already existing expertise in your community. Focusing on strengths zooms into the specifics that are happening well and creates space for teacher reflection and interrogation of practices.

## Making Space for Teacher Reflection

Teacher reflection and introspection is crucial to noticing what is and is not happening in the classroom to support each student's success. True reflection requires creating a space to be vulnerable with oneself and colleagues to dive deeper into our learnings and to recognize when changes are necessary to reach the vision of equitable mathematics instruction. *How is your professional development making space for teacher reflection and introspection related to the equity-based vision you are working toward?*

For equity to take place, it is essential that teacher reflection includes interrogating how societal characteristics play out in your system. This means starting by discussing our own identities. For professional development to create the space for teachers to dive into these ideas, where is the space for facilitators of your professional development to interrogate their own identities? How are they attending to positionality, power, and status when developing and facilitating the professional learning opportunities?

While participants in professional developments change, it is important for those designing and leading professional developments to consider the community that will be joining you, their strengths, and possible areas of growth. Professional learning opportunities can give the space for teacher reflection. You may have teachers considering and discussing what they noticed in a lesson for a question like: *What structures and/or teacher moves supported equitable participation?* Or having teacher leaders or department chairs reading literature and considering an activity where leaders make sense of whether a teaching practice is or is not supporting each of our premises. Similar to how we want students to make sense of mathematics, we want teachers to make sense of instructional practices in a way that helps them recognize which of these might offer better support for all students.

It is essential that all educators take the time to reflect on their own identities and how it connects to the world to support us as educators to make sense of supporting all students, consider who is and is not participating in classes, and notice how outcomes might fall along particular status characteristics. It can be hard for a teacher to look at the participation patterns in their class and notice, for example, that the few white students in the room are the same few students who share out. Or if someone points out that many of your girls and particularly girls of color do not seem to be included in small group conversations. Teachers noticing this for the first time might feel defensive or place the blame on the students. As educators who are a part of societal norms that stereotypically tell us who is smart and capable, who might be *trouble* and who fits typical beauty standards, it is important for all teachers to consider their own identities, privileges, and disadvantages, both by perceptions and by realities. One learning activity the San Francisco central team did together was to draw a picture of yourself, one as a perception that people might think of us as we go into school sites, and one as a reality of who we are walking into sites. For Angela, one of the authors of this book, she is Mexican American with dark hair, dark eyes, and brown skin. She knows that she taught high school mathematics for many years and has a degree in mathematics. Her drawing of the perception people sometimes have of her at sites included parent, paraprofessional, teacher's aide, and even student teacher. She never thought anything about these perceptions, never felt less-than for being mistaken in one of these roles, because there are many adults who walk into school buildings. Yet, some colleagues were surprised that these were actual perceptions she encountered. This activity supported her to recognize how, as a brown Latina, she is also changing the perception of what it means to be smart in mathematics for students. The experience also helped her always search for the strengths every individual has and notice when her assumptions of colleagues and students are wrong. We are not perfect, and so taking the time in many ways to dive into this context is a crucial first step to undoing your own biases in the world.

## Making Connections

It is essential that leaders and those providing professional development and support make as many connections for teachers as possible. *How are your professional learning opportunities intentionally making connections for teachers?* Facilitators of professional development can help teachers make connections

- **across the mathematics content and pedagogical strategies K–12;**

- **across professional learning opportunities** to enhance and continue the learning for teachers between various programs to help with similar messaging;

- **to research** and literature in the fields of mathematics education; and

- **to the community while developing and adjusting** professional development.

### Connections Across the Mathematics and Pedagogical Strategies K–12

Professional development can support teachers in making connections across content and pedagogical strategies K–12. In a detracked mathematics program, it is important for students to see connections across their K–12 schooling about what is valued in mathematics, including similar classroom structures and routines noticed by students across their school years and similar values around what it means to do and learn mathematics. Students' mathematics experiences should not be based on the teacher they are scheduled with that year. We want all students to know that they are a person who is smart

> *Students' mathematics experiences should not be based on the teacher they are scheduled with that year.*

in mathematics, who has a ton to offer, and who can learn more when doing math with others. The San Francisco team used a common curriculum with the same unit design and signature pedagogical strategies across K–12 (see Chapter 5 for further discussion on curriculum), which became the basis of much of their professional development. This created common language across the district and across site teams so if and when a teacher changed schools or grade levels, they had the tools that allowed them to jump into collaboration, content learning, and getting to know their students. Students moving from year to year would experience similar routines, like a Math Talk or a 3-Read Strategy, and be used to sharing their thinking out loud. In elementary grades, professional development can make connections across content areas, such as supporting teachers to see how talk moves can be consistent across the day, or how a language objective in math can reinforce the reading standards for the grade level. Time and time again, teachers would comment on how building groupwork and discourse routines became easier over time because this was not new for students. Professional development can also include vertical alignment activities such as examining a progression of standards and topics across grade levels where teachers read standards, do math together from activities to represent these standards, and discuss the similarities and differences to help teachers better understand how to support students and provide the cognitive demand relative to that grade level.

## Connections Across Professional Learning Opportunities

Teachers are asked to attend various professional developments throughout the year. Designers of professional development can design with connections in mind across professional learning opportunities. In a program centered on reculturing math departments with complex instruction, Jilk and O'Connell (2014) call their connection across professional learning spaces *mutually informed learning spaces* (pp. 211–212). In this model, facilitators of professional development are helping teachers make connections from one learning opportunity to the other, as well as using their knowledge of teachers and teams to inform their planning of all support provided. The San Francisco team took their experience with the Complex Instruction program in their high schools to their approach with other professional developments. For instance, when working with department chairs or other teacher leaders after a Collaboration Day, the facilitators might reference the Lesson Planning Process they used to encourage the department chairs to continue to bring this back as a reminder of a resource to use with their course teams. The facilitation team might also connect to departments leading professional developments on supporting multilingual students, allowing the math facilitation team to make connections to English Language Development Standards and particular language strategies that teachers are asked to use. Even using the strengths, interests, and challenges of teachers that you learn about with individual coaching can support stronger connections to professional development, thus supporting teacher development in a coherent way.

It is important to connect your professional development to current research on mathematics education. Providing research and literature to teachers in professional development is a way to continue to have teachers make sense of the vision and a way to show that these ideas are not only those from the central mathematics team, but rather from other knowledgeable experts across the field. Not sure where to start? Consider looking at resources available from national and state leaders in mathematics education organizations. Many organizations, such as the National Council of Teachers of Mathematics (NCTM) and the NCSM: Leadership in Mathematics Education, have position papers, books, and research briefs all based on research. Organizations such as TODOS: Mathematics for All and the Benjamin Banneker Association provide a lens on equity in their position statements and webinars that allow you to connect to math education researchers also focused on equity. Here are a few of the literature and resources connected to research that the San Francisco math team has had educators read and discuss in professional development over the years:

- The American Educational Research Association Presidential Address by Deborah Ball (2018), "Just Dreams and Imperatives: The Power of Teaching in the Struggle for Public Education"

- NCTM's (2014) *Principles to Action: Ensuring Mathematical Success for All*

- *Rehumanizing mathematics for black, indigenous, and Latinx students*, edited by Goffney, Gutiérrez, and Boston (2018).

- Smith & Stein's (2018) *5 Practices for Orchestrating Productive Mathematics Discussions*

- TODOS position statement (2020) "The Mo(ve)ment to Prioritize Antiracist Mathematics Planning for This and Every School Year"

*Connections to Community While Developing and Adjusting Professional Development*

Learn from previous professional development programs and adjust to the strengths and needs of your community. All our contexts differ and are constantly changing. Consider your community, their strengths, and challenges, and be flexible in adapting structures and content to them. Even a tried-and-true professional learning structure or activity with one group or in one year may not meet the strengths and needs of another group or different year. Here are a few examples of adjusting professional development while attending to the strengths and needs of a community:

- You previously gathered teachers by whole department or all teachers of math at a site to consider the pedagogy of instruction. However, with a new group of schools, the teachers are more connected to the mathematical content they teach and have a hard time adapting the instructional moves to their context when experiencing a math task not from their course. Therefore, adapting professional development to focus on course content with teachers attending by course teams might better meet their interests.

- Previously, release days for teachers were easier to organize in a time with a more sizable substitute pool, and in a time of substitute shortage you might take the big ideas of a full release day focused on team collaboration to an after-school time that still allowed for modeling of doing math together and facilitated time for teachers to plan together.

- You had a structure of teacher leaders joining from every middle school across the district and wanted to move this structure into the elementary grades. Yet when examining the number of elementary schools in your system, physically bringing such a large number of teachers together and paying for their time is impossible. Thus, designing a district teacher leadership model with professional development for leaders who share their learning in a public virtual space is a creative way to support many schools with the capacity and resources you have.

## LESSON LEARNED

As you develop and adjust professional learning opportunities for educators, continue to support the building of strong sustainable learning communities by building off the strengths and challenges of your learning communities, previous programs, and previous activities you created for professional development.

 **Questions to Consider for Your Context**

▶ As a leadership team, return to the questions laid out at the beginning of this chapter. Take time to consider how these can be used to strengthen or begin a professional development model that will support a heterogenous mathematics program.

- **Vision of a Mathematics Classroom:** *What is the vision for an equity-based mathematics classroom that you are hoping your students will experience in your mathematics program?* How are you providing opportunities for teachers to understand what this vision looks like and sounds like for all students?

- **Vision of the Adult Learning Community:** *How is your professional development supporting the vision of a powerful adult learning community that you want to see at your site(s)?* In what ways could you strengthen the support for teams of teachers to continue the work of collaboration and learning between or after professional development?

- **Working From Strengths:** *How is your professional development recognizing and building on the strengths of your students and teachers?* How are you working with teachers to understand the many ways we can be smart in approaching mathematics?

- **Teacher Reflection:** *How is your professional development making space for teacher reflection and introspection related to the equity-based vision you are working toward?* How are you including work to interrogate our identities (i.e., racial, cultural, languages, gender, gender identity, and sexual orientation) in the world and how that impacts our work?

- **Making Connections:** *How are your professional learning opportunities intentionally making connections for teachers?* How are teachers learning about the content that students learn before and after their grade level so that teachers can help students make connections and the teachers can support students to learn the grade-level content? How are different facilitators, who might see the same teachers in different contexts, connected so that there are collective messages relative to your visions rather than conflicting ones? How are you making connections for teachers to the many professional learning opportunities and messages that teachers receive?

## Activity 6: Lesson-Planning Activity for Teams of Teachers

▶ The San Francisco team created unit and lesson-planning templates and protocols for teams of teachers to use. The following activity is one for a grade-level or course-alike team to do together. This process will support a team to have a lesson planned with understanding the mathematics that you want students to go through, anticipating possible student approaches to the mathematics, consider where students might get stuck and barriers to participation that might occur, and then finish the process considering participation structures that can support equitable participation and access to the grade-level content. This process, while laid out in an order, might be an iterative process. As a team moves through this process, they may need to return to the math task to edit what you are asking students to do, adapting the task to their students' strengths and needs. To begin, decide as a team what mathematical activity, lesson, or problem you will work on together.

*(Continued)*

*(Continued)*

## LESSON-PLANNING PROCESS

1. **Do the math together using participation norms:** Put away our teacher hats and consider the content as learners (even if we've done it before, seek to see the problem from a different way).

   a. Norms: Talk together out loud, share strategies, approach from student strategies, work on the same problem at the same time, find a middle space to work in together.

2. **Understand the multiple abilities and learning objectives:**

   a. Make a list of the smart things we did (or *particular students* might do) as we did the math.

   b. What are the learning objectives of the lesson? What are the language objectives? What prior knowledge might our students use? Where might our students get stuck? What are the big ideas that this task will support for students in learning?

   c. Adapt your task to make it more groupworthy if needed. Find ways to open the math task so there are more ways to be smart and go about solving the math problem.

3. **Envision equitable participation and potential barriers:**

   a. What should it look and sound like for our students to learn as they do this math?

   b. What might get in the way of it looking and sounding like this?

4. **Choose structures to support equitable participation:**

   a. How can we structure the task to provide opportunities for *particular students* to show their strengths?

   b. What can we do to provide access and rigor for all students? Consider lesson structures, teacher moves, task adjustments, manipulatives, debrief of the lesson, and language development.

Source: SFUSD Math Complex Instruction Program, Revised 2021

# CHAPTER 7

·····························

# COACHING TEACHERS TO SUPPORT HETEROGENEOUS CLASSROOMS

**M**ichelle Cody, a 6th-grade mathematics teacher, sat down with one of the authors of this book to chat about her experiences with coaching. She has participated in coaching in San Francisco for about 8 years. Over this time, she has taught in two different schools and worked with a few different coaches from the central mathematics team. She shared how all her coaching relationships were special because "a true impactful coaching relationship makes it less about the coach and more about what the coachee needs." She described how one coach helped her with needs of clearing her teaching credential while another helped her dig deep to really learn about what it means to create social justice lessons in mathematics. These are lessons in which students use mathematics to explore and understand issues of social justice. During one year, Michelle was approached by her coach to want to learn more about the social justice lessons she had created, and they created a small learning group to learn about and build from the lessons that Michelle had previously created. Michelle describes her coach as learning alongside her in their conversations and time together.

Michelle also mentions how helpful it was that she and all her coaches had taken the same summer course on Complex Instruction and groupwork and how this experience was so foundational to their coaching conversations, giving them common language and areas of teaching to investigate related to what is important with mathematics and how to best support students. "I have a lot of math phobias," Michelle said. "As a Black woman who didn't study math, I found myself shrinking in the room or feeling unsure about what I know; I really was very privileged that the relationships with coaching were humanizing, supportive, loving, and all the things that I needed to be the best version of myself that I could be."

Michelle received continuous math coaching support. She described how there would be times that she would debrief with her coach, and she would *start by saying, "Ok, here are all the things that went wrong . . .," and then her coach would add, "And here are all the strengths I noticed from your students and your class . . .," or "Let me tell you about Table 7 and the conversation they were having," or "I saw Table 8 pull out their math notebooks to look back at past assignments. They really know that their notebook is an important resource." Michelle described, "It was always so helpful to have a coach, someone you planned the lesson with, come into your classroom and scribe and later help you interpret and create the space for you to reflect later. As a teacher, you have things in your mind that you want students to say and feel and do in math class and then sometimes during class our students are very overwhelmed learners and their interactions pull you to hear things you don't want to hear, the negatives, so having a coach having the job of seeing the strengths, they are able to help realign your mindset with real data and feedback. They let you know that you are making an impact, and this is working in some places so then we can think about 'How can I get this to be institutionalized into more of my practice?'"*

*Michelle was one of many middle school teachers who experienced the change from tracked mathematics classes to heterogeneous classes. She remembers having general education classes and then honors classes and remembers how students in the general education classes perceived themselves as less capable than the honors students because of how the structure of math classes was presented to students. As someone who did not grow up believing she was a "math" person, conversations and learning in both coaching and professional development have supported Michelle with opportunities to question what she was thinking before and now she is able to bring this experience to her students to help them all realize that they are in fact math people. Michelle finished by saying, "Coaching, I believe, is super impactful, important, and transformative to teachers being able to have a thought partner through the process to help make sure the students are getting quality education and a quality learning experience."*

## WHAT TO EXPECT IN THIS CHAPTER

*In a detracked system, teachers are being asked to provide a heterogeneous classroom experience where all students have access to rigorous grade-level mathematics. Creating these discourse-rich, equity-centered classrooms that bring together students who are neurodiverse, whose primary language is not English, who have different cultural backgrounds, who bring a variety of ways of thinking, who may not remember certain mathematics content from previous grades, and who are currently thriving in a unit of study is a complex task. Without clear support for teachers who are asked to create successful heterogeneous mathematics classes, the same inequities will continue to exist, even in detracked classes. A 2015 research brief from the National Council of Teachers of Mathematics (NCTM) discusses the increase in research of mathematics coaching and how coaching one-on-one and in group settings has shown to improve both teacher practice and student achievement. Coaching can provide another pair of eyes to observe and a thought partner for reflection to improve practice and show impact on student learning. Coaching becomes an essential part of partnering with teachers to work toward that equitable classroom.*

> Without clear support for teachers who are asked to create successful heterogeneous mathematics classes, the same inequities will continue to exist, even in detracked classes.

*While coaching is a type of professional learning opportunity, this chapter expands on how coaching should be used as an integral part of a systemic approach to supporting heterogeneous classrooms. In this chapter, you will see many connections to the big ideas in the last chapter related to professional development. Coaching should*

- *Be centered on a vision for an equity-based mathematics classroom as well as a vision for a strong learning community*

- *Build from the strengths of teachers and students*

- *Make space for reflection and introspection*

- *Support teachers to make connections*

*This chapter continues to reference these big ideas as essential throughout our discussion of coaching structures and approaches.*

*While coaching can occur with individual teachers, it can also occur with teams of teachers and teacher leaders. In this chapter, when referring to coaching, we often refer to the one-on-one coaching between a coach and an individual teacher, including a pre-conversation, an observation, and a post-conversation, or a combination of these. When discussing the coaching of teams or leaders, we emphasize that difference.*

## QUESTIONS TO CONSIDER WHILE READING THIS CHAPTER

As you consider how coaching can support your system, we hope you also investigate the following questions as you are designing coaching support for teachers:

- **An Approach to Coaching That Is Strengths-Based and Adaptive:** What is the approach to coaching that you would like coaches to take in sessions with teachers in your system?

- **Attending to Equity in Coaching:** How are your coaches attending to equity both in the classroom and in conversations with teachers?

- **Deciding Where to Focus Coaching:** What are the strengths, key leverage areas, and areas of need to help make decisions of where to best focus coaching to build a strong and sustainable adult learning community that is working toward practices to support equitable outcomes for all students?

- **Supporting Coaches:** How is support for coaches built into your structures so coaches learn content, pedagogical strategies, and better understand your approach to coaching?

## BIG IDEAS OF DESIGNING COACHING SUPPORT

As you saw in the vignette in the beginning of this chapter, Michelle describes her experiences with coaching. She describes having supportive coaches who focused on her strengths and the strengths of her students, who adapted to her needs, attended to equity, and learned alongside her. Additionally, Michelle received coaching because she fell into a variety of focal areas over the years. Her coaches also participated in a variety of learning opportunities to support a common approach to coaching while allowing each of them to attend to each of their unique identities in relation to those of hers and her students. To help understand the big ideas of coaching, the following chapter uses examples from Michelle's story, stories from additional coaches, and stories from those leading coaches in the San Francisco Unified School District (SFUSD). The San Francisco math team credits much of their work around coaching, the approach to coaching they used, and the big ideas and lessons learned around coaching to the Complex Instruction work happening in the district at the inception of this work with the consultants named in the vignette of Chapter 9 and the collaborative learning the team has done building from this work over the years.

### An Approach to Coaching That Is Strengths-Based and Adaptive

Similar to helping teachers experience for themselves what an equity-centered mathematics classroom is like in professional development, you want your coaches to take a similar approach to coaching your teachers. *What is the approach to coaching that you would like coaches to take in coaching sessions with teachers in your system?* The San Francisco team took a strengths-based and adaptive approach to coaching that they learned from the consultants who supported the implementation of Complex Instruction at some sites (Alloway & Jilk, 2010; Baldinger, 2014; Jilk 2016). Together, they worked for many years to better understand this approach and how it can be useful with teachers. This strengths-based approach came from the similar approach to working with students, noticing and building from teacher and student strengths in the classrooms, understanding who they are as individuals in the world and their coaching relationships, to then adapt the coaching moves specific to each teacher.

### Noticing Teacher Strengths

In Chapter 6, we laid out the premise that *all teachers have strengths in teaching and learning.* Coaches must work to notice the strengths teachers bring, because this does not often come naturally to most people. What constitutes a strength in teaching and learning? To support teachers to reach the vision of an equity-based classroom, a coach must notice strengths that connect both to the larger principles of teaching and learning and find qualities from the vision they want to see across all classrooms. Coaches can look for evidence of a teacher's strengths in many places, including their teacher moves, mathematical task design, the resources on the walls of their classrooms, the structures they use within a lesson, how students see themselves as learners of mathematics, and from things the coach hears students and teachers say in the classrooms and in coaching conversations. Strengths of students working in the classroom can be connected back as the teacher's strengths as a result of something the teacher set up. For example, Angela, one of the authors of this book, once was in a classroom where middle school students were sharing out in small groups and the whole class was using language that seemed to mimic the same sentence frames for discussion over and over. As she noticed the pattern, she knew this was a strength that the teacher must have created and supported over time, so much so that without prompting, students used this language naturally. When looking around the room, Angela noticed clear posters with the exact language students were using! This very important strength was one that she could bring up in the debrief.

### Adapting Coaching Moves to Who the Teacher Is

As a coach, it is important to build a trusting relationship with each teacher. Each teacher is different, which is why the strengths-based and adaptive coaching is not a one-size-fits-all model to coaching. Additional strengths of teachers become apparent in who each teacher is as a person, including their many identities, their background, and their interests. In Michelle's discussion, you could hear her describe a humanizing and supportive experience with each of her coaches. Each coaching relationship was similar in that they built from her strengths, adapted to her current needs and questions, and worked toward a similar vision; yet they were also very different given their various backgrounds, the length of time of their coaching relationships, and each coach's various strengths. One extremely important caveat to remember as a coach is to *not* assume a particular background on a teacher based on stereotypes and biases that we each hold. For instance, do not assume someone had a tough or poor upbringing based on their race. Or do not always assume that if you share the same culture or race that you also share a similar background. Instead, be curious and give space for teachers to tell you about who they are, their relationship to mathematics, why they are teaching, who they are as humans, and so on.

As you coach, you can support teachers to notice their own strengths and consider a next step in their practice. The classroom is an ever-changing place. The art of teaching is all about considering strategies and ideas for the classroom, seeing how those ideas impact students in participating and learning, and then revising your ideas for future lessons. Building your teacher toolkit with ideas and strategies takes years; a coaching relationship can support this process by adding eyes and ears to notice strengths that a teacher may not otherwise notice, support reflection, and consider new ideas based on the teacher's strengths and interests. While the coach is taking note of all the strengths they notice in a classroom and over time, they can be strategic about how and where in a conversation to name these strengths. One debrief might look like having the teacher start by sharing strengths they noticed in the lesson and what they observed from students, allowing the coach to learn more about what the teacher gravitates toward in noticing and wondering about in a lesson. The coach can add additional strengths and make connections regarding how these specific strengths support the principles of teaching and learning in relationship to the district's goals and vision. For instance, in a coaching interaction a coach might say, "By coming around to each group and asking them to justify, you were holding them to high expectations." A teacher, however, might respond, "I didn't know I was doing that!" This approach to coaching supports the teacher's growth so that they will be able to notice, reflect, and make these connections to the vision on their own and with their colleagues. Additionally, a coach might make suggestions based on a teacher's particular strength area or areas of interest. For instance, if together a teacher and coach are noticing some inequitable participation among students, they might suggest a task design change for the teacher who thinks deeply about content, or they might suggest emphasizing a particular norm or student role for the teacher who gravitates toward thinking about norms in their class.

## LESSON LEARNED

Focusing on strengths and adapting to a teacher means a coach needs to really be curious about who a teacher is and help them develop the tools to reflect on students' learning and consider possible next steps. This is the opposite of a coach visiting a class with a lens of how can I *fix* this lesson, often offering similar ideas to what they would have done in that situation.

## Attending to Equity in Coaching

It is critical to pay attention to equity while coaching from all angles. This includes the classroom dynamics among students, interactions between teacher and students, among teams of teachers, in coaching relationships, and among your own coaching team. For the San Francisco team, learning about how status can influence participation and learning (Cohen & Lotan, 2014) has greatly influenced them to consider the impact status can have in all relationships. With this work, they consider how status characteristics, like race, gender, and other characteristics described in Chapter 6, might impact participation and learning together. *How are your coaches attending to equity both in the classroom and in conversations with teachers?*

It is important to build relationships with teachers to be able to work toward conversations where we think about who is participating in small groups and who is not. Considering the deficit narratives out there about students of color, it is essential that coaches are always on the lookout to notice and share with teachers the mathematical strengths they see from students of color in the room, helping teachers expand how they see their students as smart. It is also important to use evidence to share when the groupwork is not quite working yet and how students may not yet be accessing the conversation or materials. For instance, in a coaching conversation where the coach captured quotes of a small group, they discussed the evidence together pointing out strengths and patterns of participation. "At Table 8, two boys on the left side of the table were talking about the math, but the girl across from them wasn't doing much of the talking or pointing to the task. And then, after you came over to the group and posed that tough question to them, the boys were grappling with it and then asked the girl what she did and she shared. That is exactly why we want challenging problems that make them need multiple ideas, and in this case they needed her to help them solve it." Using direct evidence such as quotes or even video allows the coach and teacher to discuss the evidence together and use it to make sense of students' learning. They could follow up this noticing to ask the question, "So how can we get these boys to continually recognize her as having something smart to offer?" and then brainstorm ideas to support this question.

Another strategy used to help teachers and coaches visually see how a group of students is participating is to map the students' participation with a diagram (Featherstone et al., 2011, p. 95). While this diagram is not limited to coaching, some of the San Francisco coaches brought this approach to their coaching with teachers. Figure 7.1 shows a few different mapping conversations that could be captured by a coach or teacher to show who is talking to whom in groups.

**FIGURE 7.1  PARTICIPATION DIAGRAMS**

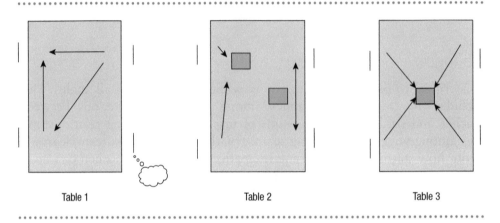

| Table 1 | Table 2 | Table 3 |

Figure 7.1 shows a few different small groups of four. Table 1 shows the student in the top right talking to both students across from them but not to the student next to them. The pair on the left side of the table has an interaction that is only one directional, and the student in the bottom right corner is looking on and thinking to themselves, but not sharing their thinking out loud. Table 2 shows a pair of students sitting on the left side of the table, conversing with each other, and the materials are more closely centered in front of the student in the top left. The pair on the right side of the table seem to be discussing equitably with their materials centered between the two of them. Neither of these pairs at Table 2 are conversing with each other. The diagram for Table 3 shows the type of visual that represents the vision of what we want happening at a table, with the group having one conversation, each student participating with the group sharing their thinking out loud, and the materials centered in the middle space. Using these types of diagrams with a teacher in a coaching session, coupled with the teacher's knowledge of students and their collective noticings of strengths and positionality in the room, they can conjecture about why the pattern of participation is happening the way it is and reflect on ways that might support more equitable participation in the groups. This, along with a trusting relationship, allows a teacher and coach to recognize how characteristics like race and gender might be playing out in a classroom.

Attending to equity in coaching also refers to attending to a coach's own identities, as they relate to students in the room and to the teacher. The coach needs to examine how their position can shift due to status and power and be mindful of how these might alter working with teachers of a different race, ethnic background, age, and gender from their own. For instance, a coach who is older than a teacher may want to enter a room and when introducing themself to students, say, "I am (first name), I am also a math teacher, and I am so lucky I get to learn with other math teachers and listen to your strengths today." This type of introduction can help a teacher feel at ease by reiterating the learning-together aspect of coaching versus a fixing or evaluative coaching stance. It can also help minimize the power often automatically given to a coach walking in, perhaps unintentionally undermining a new teacher or teacher of color. In another coaching instance, with a coach who is a younger woman of color, it is important for all students to see a woman of color in a leadership role in mathematics.

## LESSON LEARNED

The work in coaching conversations and relationship building is complex, which includes the many layers of our identities and experiences, related to who we are as people, our cultures, who we are as learners, and who we are in relation to mathematics. A coach needs to be able to navigate the feelings of (in)security that we all bring and that can resurface as conversations emerge of who is and is not participating in a class.

### Deciding Where to Focus Coaching

At the systems level, it is important to consider how and when in your system coaching for mathematics teachers will happen. It is nearly impossible to have the funding for ongoing coaching for every teacher. Michelle was paired with a coach for a variety of reasons. At times it was because she was participating in a central math program that included professional development and coaching for interested school teams; for a few years it was through induction support for teachers clearing their credential, other times it was because she was a middle grade teacher and every middle school received a coach at the beginning of the district's detracking policy, and other times she was at a focal school with a high percentage of students who have not experienced much success in mathematics. Therefore, it is important for leadership to think about where the focus of coaching might best serve your system. If you are a small school, you may start by focusing on all mathematics teachers. However, if you do not have the luxury to have all teachers coached, *what are the strengths, key leverage areas, and areas of need to help make decisions of where to best focus coaching to build a strong and sustainable adult learning community that is working toward practices to support equitable outcomes for all students?* Let's explore several possibilities:

- Focusing on schools with those most underserved
- Coaching teachers who are open and committed to learning
- Building critical mass by focusing coaching on coaching teams
- Connecting to additional coaching in your system

### Focusing on Schools With Those Most Underserved

When deciding where to focus coaching, it is important to start by considering the students and groups of students who have been most underserved in your system. Examine your system and note where there are students having rich learning experiences and note sites or classrooms that could use the extra support in particular areas, such as a site with a grade level with two new teachers, a long-term substitute teacher, or a high population of multilingual learners with interrupted formal education. SFUSD, for example, is a large urban district with over 120 schools. The district population is extremely racially, culturally, and socioeconomically diverse, and there are schools with higher percentages than the district average of groups of students who have been most underserved in the district. Michelle, the teacher you read about in the vignette, was a founding teacher at a new middle school that opened in 2015. This school has a focus of providing STEM opportunities for students while drawing from the richness of their diverse community. Teachers and staff at this school have also been working to mitigate the complex issues of poverty and teacher turnover that highly impact the learning experiences of students. This school has some truly dedicated educators who have worked hard to create a community where students and their families thrive. This school has also continued to receive centralized support, including math support through coaching. Focusing coaching in schools with populations that have traditionally been underserved must come from a strengths-based perspective and not from a deficit mindset about the school. In particular, it is important for coaches to bring a lens of strengths to sites where teachers and students have both experienced trauma in mathematics and where their previous experiences led them to see deficits of themselves and their students when it comes to mathematics. Michelle speaks to how important it was for her to have a coach who would bring a strengths-based

> *Focusing coaching in schools with populations that have traditionally been underserved must come from a strengths-based perspective and not from a deficit mindset about the school.*

approach to their conversations, the impact it had of showing her the strengths that she had not noticed of her students, and how it helped her see that she had practices set up to create these experiences for her students. Being in a community in which society has conditioned people to view particular students as not smart at math can cause teachers, with the utmost intent of support, to over-scaffold math problems or to provide mathematics tasks that are well below grade level. Coaching can provide an extra pair of eyes where coaches and teachers can learn together how best to provide tasks that hold high cognitive demand with the grade-level mathematics while providing access through open tasks, culturally relevant pedagogy, and explicit understanding of the mathematics.

Coaching that centers teacher reflection on the individual while supporting the whole learning community in their growth can provide the support for teachers and classrooms that might be struggling. The previous chapter discussed how strong learning communities can sustain this work, with teachers continuing to learn together and staying at their site for years. When teachers find support that provides a working environment where they can feel the wins in the classroom and feel they are a part of a team that supports their growth, teacher stress goes down, teacher turnover lessens, and students have a more consistent and safe learning environment. Thus, focusing your coaching on sites and programs with a higher need using a strengths-based approach is essential for a district, especially when the coaches' capacity might not allow a coach at every site. When we can support schools to create a strong learning environment, we can see the richness in their mathematical strengths and the strengths of their culture and of their community.

### Focus on Coaching Teachers Who Are Open and Committed to Learning

Making changes in the classroom requires coaches and teachers committed to learning together. Learning through coaching is a vulnerable practice that requires introspection and interrogating one's practice. Thus, a trusting relationship between the coach and teacher is essential to supporting the learning process. While the coaching relationship makes space for a reflective practice, it is also helpful if the teacher is willing to dive into the strengths that support different students, to examine practices happening in their classroom, and to try new things to move their practice forward. The stronger the coaching relationship is and the more willing a teacher is to be reflective and try out new ideas, the closer a classroom can make changes toward the vision of an equitable math classroom. Michelle spoke about how having the same coach visit her classes on a frequent basis allowed for their relationship to grow. She also described how her students became used to the coach and how he became a part of their community, allowing students to be their authentic selves with him and allowing him to ask students questions about their math work in class.

Coaching can be a sustainable practice if teachers and coaches are committed to the coaching process. It is important that central administrators of coaches, coaches, site administrators, and teachers create and protect the time for coaching to happen. First, coaching should be seen as a support for teachers, as opposed to it being extra work or a straining time commitment that was put on teachers. Teachers who are committed to learning, like Michelle, will capitalize on opportunities like coaching and professional development because they know that these sessions support them in reflecting and improving their practice for the students. Structurally, finding the time for coaching can be difficult, with planning and debriefing sessions happening before or after school, lunch time, or during a prep period. During these times, teachers have many responsibilities: planning lessons, grading, attending meetings, covering lunch duty, covering an absent colleague's class, supervising a student club, and so on. Thus, buy-in for teachers to use these times for coaching is not an easy feat, unless it can be understood by all as a helpful time to have a thought partner for the planning and reflecting they are already doing. Teachers committed to coaching find ways to make sure their coaching schedule happens. This may mean working to protect a planning meeting with a coach by letting the site's administrative assistant know they need to protect their prep on a particular date or letting students know that tutoring cannot happen after school because they have a meeting. Coaches can also help teachers advocate for their protected time by sending a proposed coaching schedule in advance, encouraging teachers to reply with alternative options if needed. Furthermore, those coaching often have additional job responsibilities such as providing professional development and/or curriculum support. It is important for administrators and coaches to prioritize the coaching work. This might mean if a central team makes a commitment for their content specialists to coach six teachers five times a year, then the administrator encourages the coaches to block time on their calendar throughout the year to ensure coaching happens in addition to other responsibilities.

As much as possible, it is important for those who create the pairings of coaches and teachers to adjust and adapt in ways that are best for supporting the ongoing learning for teachers. Coaching can come across roadblocks without a trusting learning environment or a commitment to the coaching process. For example, if a coach is reassigned the following year to a different set of teachers, is there a way to keep existing successful relationships together so that the previous teacher does not have to start a new coaching relationship? Or if a coach and teacher seem to have not yet built a trusting and learning relationship, with a teacher having a hard time diving into the coaching relationship due to perceptions of competence (from either the coach or teacher), is there another option for a coach where the pairing might be successful? Or if a teacher or coach continues to cancel coaching sessions while prioritizing other work, is there another teacher in your system who would be welcoming of coaching resources? Ultimately, the more your program can commit to deep coaching, work with teachers who are open to coaching, and adapt to better pairing of teachers and coaches, the more trusting coaching relationships can be developed to help the work move closer toward your vision.

## Building Critical Mass by Focusing on Coaching Teams

To make systemic change requires creating a critical mass of teachers working collectively toward equitable change so that the experiences in mathematics for all students improve and where identifiers such as home language, race, ethnicity, and disability are no longer predictors of success in mathematics. We can strategically coach teachers who are committed to working toward your district's vision and who are open to collaborating with colleagues, with the hopes of building strong site learning communities. Without a critical mass of teachers working together to provide similar equitable experiences for students, our students are left with different beliefs about who can do mathematics and what mathematics is, and the same inequities will persist. Thus, coaching should be seen as a structure to develop teacher practice and the development of strong learning communities, both at the district and site levels that will stand the test of time.

When supporting a school for the first time coaching the full department may not be possible, look for small teams of teachers to begin your work. You may choose to start with the grade-level team most interested, with those who are already making movement toward your vision, or with a teacher who is influential to the larger team. This choice is complex. For example, a middle school math department has the entire 7th-grade team of three teachers, one 6th-grade teacher, and one 8th-grade teacher willing to commit to coaching and professional development. Considering the learning community as the unit of change, we want to encourage teachers to regularly talk with each other, plan together, and brainstorm how to use new strategies they are learning. Some might wonder, do we coach all interested teachers? When deciding who to coach, think: Who will the 6th- and 8th-grade teachers plan with? If these two teachers plan together in a small school design, maybe they can be considered their own team and coaching them makes sense, including team coaching to help them see what planning across grade levels can look like. However, if these two teachers are part of separate grade-level teams, their other colleagues are not trying the same strategies, and they do not have the structure to plan together, they may not have the daily support needed to sustain the change to their classrooms. Thus, a decision might be to start with the 7th-grade team this year, both individual coaching and team coaching, such as planning or peer observations. The excitement and successes of the 7th-grade team can help the other teachers find a grade-level partner for coaching the following year.

As much as possible, it is important to include teachers in the conversation when recruiting for new coaching relationships, to gauge excitement and the teacher's and team's general areas of inquiry. If a teacher makes an assumption that a principal assigned them coaching because their classroom is *bad*, then the coaching relationship could start with the teacher in a defensive mindset, making it tough to build that trusting relationship. In contrast, coming from a strengths-based perspective, finding teachers who are willing to collaborate and are open to learning are strengths in a community that becomes the spark to build from. The same ideas apply when considering which schools to involve in coaching early on. Start the deep coaching work with site departments that have a large amount of interest by teachers and are already making movements toward equitable instruction, building strong classrooms to show the vision you are working toward is possible and sustainable. Be sure to support these teams so that these strong learning communities should be able to continue this vision past teacher turnover and beyond coaching. As you work to expand, lean on the expertise and excitement from your early adopters to support additional teacher teams and schools in future years.

When starting to make change at a large scale, such as a detracking policy, the implementation of a new curriculum, and/or wanting to see change in instruction toward more student-centered classrooms, the early adopters into the work can get the momentum going. A critical mass of teachers at a district level might be math departments from half of your schools who participated in a previous professional development and coaching program and are implementing an equity pedagogy that centers student discourse in their classrooms. In San Francisco, they have seen early adopters like these turn into the critical mass. These teachers and site teams can speak to the strengths of heterogeneous classrooms, advocate for the district detracking policy, be a part of the district curriculum writing or adoption groups, and become leaders across the system as teacher leaders, coaches, and administrators. At a site level, a critical mass might be that four of the five math teachers plus a few special education math co-teachers from your school have participated in professional development and all these teachers are willing to engage with coaching this year, making learning together and similar classroom practices as the norm for your students. Considering the department as the learning community that is the unit of change (at elementary this may be a grade-level span of teachers), critical masses should be seen as a majority that is tipping the scale toward the change described by your vision, changing teaching practices, and increased learning outcomes for all students.

Additional areas to focus your coaching support can be intentional based on critical junctures in students learning trajectories or particular structures happening in your context. There may be particular needs at certain grade levels. For example, you could focus coaching where there are

- Foundational courses students take when they start at your school, such as Math 6 in a Grades 6–8 middle school or Math 1 or Algebra 1 for students in 9th grade at the high school level

- Bridge years, such as eighth or 9th grade, that show critical data points that predict whether students are on track for high school graduation

- Large changes to a community in certain grade levels. In San Francisco, the changes to content in middle grade mathematics courses, including the titles and content of courses in eighth and 9th grade as a result of new standards and the move to detrack, led to a lot of family and teacher confusion. Thus, middle grades became a focal area to provide coaching support at the launch of the detracked course pathway policy

Knowing your teachers, sites, and the support they are getting currently is also critical. Say for example that you have decided that the central math team has decided to focus coaching on 8th grade at all schools, yet the central mathematics team had previously been supporting the 7th-grade team for 1 year prior, building relationships with these teachers who became increasingly more open to change and learning. If your coaching team does not have the capacity to coach two teams at this site, dropping the 7th-grade team to bring in an 8th-grade team would be unproductive to the idea of building critical mass. For this site, it may be better to continue working with the 7th-grade team for at least a couple more years. This team can then build expertise and excitement at the site and increase curiosity from other department members. When it is time to bring in the eighth-grade team, it might even be possible to move a 7th-grade teacher to this team to distribute leadership and support growing the work across teams. Once you have your teams of teachers to coach, coaching can include one-on-one coaching cycles with individuals, coaching a course- or grade-level team through collaborative planning times, and coaching teacher leaders to consider leadership strategies that the teacher can facilitate with their team or next steps they can take to advocate for the work.

## LESSON LEARNED

One-on-one coaching, team coaching, and leadership coaching should be intentional at supporting the adult learning community, which is essential to create change. Working to recruit teachers into coaching with a course- or grade-level team or at least a site-based colleague can support building a critical mass of teachers doing the equity work, which over time can create the excitement and learning that can spread to the whole site or department.

It is important to recognize where there might be additional coaching of teachers your team wants to support with math instruction happening in your system, and it is essential to find ways to connect with these coaches about the vision for the math classroom that you are working toward. In school systems, teachers encounter coaches through various purposes including an induction coach for new teachers clearing a credential who might be another teacher at their site or from a different central office department, a site-based multiple subject coach or literacy coach at elementary, an intervention coach, a site-paid mathematics coach, outside coaches or support providers set up by grants, or district central coaches provided through other departments or programs separate from a centralized mathematics team. Coaching for new teachers and specialized areas such as for multilingual learners is essential to any system, and it is imperative to align the visions of these coaches to prevent mixed messaging to teachers and ultimately to coherently work toward the same vision. When considering how to align coaching for teachers, first consider if there is a person on the central math team who would best suit this coaching pairing, based on their expertise, strengths, and previous experience with a site or teacher. This might mean asking a different central department to pair a central math person with a particular new teacher for induction coaching because they are already working with other teachers at their site or because they too have a special education credential. In another example, a member from the central math team is coaching teachers at a school site and some of the math teachers are receiving additional coaching for supporting their sheltered math classes with all students who are emerging multilinguals and are new to this country. It is so important that these teachers are receiving this specialized support. Find ways for the central math coaches and the multilingual coach to connect, learn about each other's work and resources, and help each other understand the visions you are working with teachers to achieve. This can look like informal meetings, participating in each other's professional learning opportunities or coaching cycles, or creating new learning opportunities for teachers facilitated by the two coaches. You may realize that you have inconsistencies with how you work with teachers or how you view the classrooms, and with a strengths-based approach you can, over time, recognize the similarities and begin to make connections for yourselves and for the teachers and sites you are working with, coming to a more coherent vision of math instruction together.

## LESSON LEARNED

The more we can make connections to other coaching initiatives and professional developments that teachers are involved in, the more coaches and professional developers can create a more coherent vision of math instruction for those supporting teachers in various roles and for the teachers receiving a more coherent message. Participating in similar experiences, such as a coach attending another department's professional development with a teacher, will allow both the coach and the teacher to make connections to this experience in future conversations.

## Supporting Coaches

As you consider using coaching as a large support for sites, is there a structure set up for someone to guide the learning for the coaches? New coaches coming directly from the classroom often begin coaching without the full support of knowing *how* to coach. When walking into a classroom that someone has framed as "they need help," a new coach will most likely bring ideas to fix the issues they see, which can be problematic in a few ways. First, teachers are intentional about their practice and have reasons behind their choices. Advice might be something the teacher already considered but intentionally chose not to do for a reason. Second, all situations could have multiple approaches to *fix* it. However, an approach a coach might have used in practice based on their own strengths for a certain situation might be contradictory or unrelated to an approach a teacher has been using. Third, the suggestion from a coach might be to address an area of the lesson that is not the interest or curiosity of the teacher. For example, if a coach is focusing on how norms could help more students participate and the teacher's interest gravitate more to the mathematical content, the coaches' suggestion may not resonate. With the variety of ways coaches can approach a coaching conversation, *how is support for coaches built into your structures so coaches learn content, pedagogical strategies, and understand a similar approach to coaching connected to your vision?*

When the San Francisco team decided to focus coaching on the middle grades, there was also new funding to support a new centralized middle school coaching team as well as a new program administrator as the supervisor for this team. Ho Nguyen, one of the authors of this book, took on the role as the program administrator and recognized the need to establish the vision of a mathematics classroom for the coaching team and the vision for coaching. He facilitated the coaches' learning through doing math together, observing teachers' classrooms, watching video, reading, moving through various learning protocol activities, and setting aside ongoing meeting times to do this learning together. Toni Allen, one of the coaches who started with the coaching team that first year, describes that through these conversations and observations, as a team they "began recognizing that we were starting to see through the same eyes, the same lens." This created such a strong culture of learning among the coaches that this team then used their own time to do rich learning together. According to Toni, "We weren't coaches who were talked down to, but we were a co-learning team; we were building our own understanding of the vision of a math classroom and of coaching together."

There are additional structures that can support all coaches, including those new to a coaching team, such as apprentice coaching and co-coaching. For both structures you might see two coaches planning, observing, and debriefing with a teacher. The coaches might take turns leading the conversation while making space for the other to add, sort of similar to how two teachers might co-teach a lesson. During both, the two coaches might meet before and after these sessions to discuss a possible focus, share strengths they noticed, consider which strengths might offer some next steps, and debrief how they thought the coaching session went. In apprentice coaching, it might be strategically pairing a new coach with a coach experienced in the coaching philosophy and, while they are learning together, it is an intentional pairing to support the new coach in understanding the methodology and being successful with teachers. You might connect apprentice coaching to apprenticing a student teacher with a cooperating teacher or apprenticing a new teacher into a math department. Co-coaching might be seen as pairing two colleagues who are wanting to learn about their practice in deeper ways. This would be similar to a site team's course- or grade-level team collaboration time.

In both structures, it is important to bring the strengths-based approach to the relationship with your second coach. Being strengths-based does not assume that each person brings the same strengths, but it does recognize that each person comes with strengths. Therefore, in an apprenticeship model, it is important to recognize the expertise that a more experienced coach brings while also acknowledging and making space for the strengths a new coach brings, such as really being able to put themselves in the teacher's shoes. Having both coaches coach at the same time can be confusing, can pull the conversation in different directions, and can leave less time for the teacher to do the reflecting. Therefore, it is helpful to name who will lead the conversation, which might be different by teacher or as the year goes on as the apprenticing coach becomes more comfortable. Similar to being in the classroom, when running professional development, and between coach and teacher relationships, it is important for coaches to attend to power dynamics and possible status issues between the coaches in an apprenticing or co-coaching relationship and the teacher they are working with. To manage these possibilities, the more experienced coach should be aware of this and notice and name strengths of the co-coach, just as they would with the teacher. This special attention to the teacher's learning agenda as well as the learning of a coach being apprenticed takes an immense amount of cognitive load and will take extra time between the coaching conversations that build the relationship and establish common vision between the coaches.

## LESSON LEARNED

Learning about the vision for mathematics teaching, the vision for a learning community, and the vision for coaching is essential to your coaches' growth and the growth of your mathematics program. Having time for coaches to learn together provides more ideas for each coach to learn from one another, eventually helping the coach come to a deeper understanding of mathematics teaching and learning and coaching. These learning times also support a more coherent message on what the coaching work is trying to support across your system.

▶ Returning to the questions at the beginning of this chapter, your team can take some time to consider these questions to think about how and where coaching can be an integral part to supporting teachers with creating engaging learning environments for all students in heterogeneous classrooms.

- **An Approach to Coaching That Is Strengths-Based and Adaptive:** What is the approach to coaching that you would like coaches to take in sessions with teachers in your system? How are your coaches supported to notice the many strengths available to be able to work with teachers to build off these strengths in ways that are adaptive to each individual teacher?

- **Attending to Equity in Coaching:** How are your coaches attending to equity both in the classroom and in conversations with teachers? How are coaches recognizing the role our identities, including race and gender, can impact relationships and participation in the classroom and in coaching?

- **Deciding Where to Focus Coaching:** Where will your coaching be focused? What are the strengths, key leverage areas, and areas of need to help make decisions of where to best focus coaching, to build a strong and sustainable adult learning community that is working toward practices to support equitable outcomes for all students? Are focal schools and teams being chosen based on the most underserved groups in your system? Is your team working to build a critical mass of teachers as a site by including teachers who are open to learning and committed to coaching?

- **Supporting Coaches:** How is support for coaches built into your structures so coaches learn mathematical content, pedagogical strategies, and better understand your approach to coaching? How are you helping all coaches get clear on the vision for an equitable classroom? How are you building a strong learning community among coaches and support providers?

# Activity 7: Not "Just Like Me" Activity

▶ The following protocol is meant for a group of coaches to do together as a learning activity. The purpose is to focus on the strengths of students playing out as well as a recurring issue happening in the classroom with a collective brainstorm. Our coaching ideas are often limited based on putting a teacher into a box and trying to make the teacher "just like me." Therefore, this protocol can be used to expand our repertoire for next steps in coaching for the entire coaching team, not just the presenter, and can provide a reinforcement and additional tool for strengths-based and adaptive coaching.

---

### Not "Just Like Me"

(Approximate time: 44 minutes)

- Part 1 (7 minutes): Name the problem

  - Presenter:

    - What is happening for kids (strengths) and what is not (the recurring issue— important to name the issue)?

    - What are you seeing in the classroom (only what's happening with students, not the teacher)?

- Part 2 (10 minutes): Brainstorm possible solutions (Presenter listens only)

  - Group shares possible solutions or insights.

  - What might a teacher do to make progress on this classroom problem, to create something different for kids?

- Part 3 (5 minutes): Knowing the teacher

  - Presenter:

    - What is the teacher already doing to work on this problem?

    - What do you know about who this teacher is, what commitments they have, what their strengths are?

- Part 4 (15 minutes): Considering Next Steps

  - Group and presenter: What approaches might possibly make sense for this teacher? (The point is not to land on one but to consider what we know about the teacher.)

---

*(Continued)*

*(Continued)*

- Part 5 (7 min): Group reflection:

  ○ What are your new learnings?

  ○ How did this help you think about strengths-based and adaptive coaching?

  ○ What are you willing to try?

Source: SFUSD Math

_____

_____

_____

_____

_____

_____

_____

_____

_____

_____

# PART 3

# MAINTAINING DETRACKED MATH COURSE PRACTICES

# CHAPTER 8

......................

# CRAFTING SCHOOL SCHEDULES THAT SUPPORT HETEROGENEOUS CLASSROOMS

*I*magine a student named Alicia in a U.S. elementary school who is multi-lingual. The first language she learned as a child is Spanish. Alicia loved when her mother would tell her stories in Spanish at home and her father would read to her in Spanish. Throughout elementary school, Alicia felt challenged to learn to read and write in English. By 5th grade, Alicia was labeled a Long-Term English Learner *by the leaders in the school system because she had been in formal schooling for 5 or more years and was not able to pass an English development test. Consequently, she was referred to by school leaders using an acronym that sorts our students: Alicia was an LTEL. Alicia continued to be pulled out of class for designated English Language Development (ELD) instruction with a specialized teacher. Her inability to pass the ELD test kept her in this category and provided her with targeted instruction to support her ELD.*

*When Alicia graduated from her elementary school and was about to start middle school, she received her class schedule. She noticed she was placed in an ELD class, like her pull-out class in elementary school. Her math, science, social studies, and English courses in her schedule had this label of* sheltered *in front of them. Alicia wondered what this meant. What Alicia did not know was that a sheltered class was designed for students still learning English.*

*When Alicia showed up to the first day of middle school, she was excited and nervous. She loved seeing some of her friends from her old elementary school. Yet she felt a bit overwhelmed with how many courses she had to take. She soon noticed that many of the students in each of her classes were the same, some of whom she knew and some of whom were new to her. Some of Alicia's friends from her old elementary school had other types of* courses in their schedule that did not have the word sheltered in front of them. Unfortunately, Alicia saw less of these friends because she did not overlap with any of them in her class schedule.

As the year progressed, Alicia would learn that the students in the same classes as her would be with her throughout the school year. Many of these students shared the same category as Alicia—that is, a student bureaucratically classified as LTEL. Like Alicia, year after year, many of these students struggled to pass an English language test, which kept them in these specialized sets of courses, thereby limiting their access to some courses throughout middle school. Yet, were these courses in sheltered English in fact designed to support the language needs of students like Alicia? Could something designed to be helpful in fact hinder Alicia's access to class content?

Alicia started to accelerate her learning and interest in her sheltered math class, so she approached her teacher about changing to a more challenging math class in her second semester. Alicia knew some of her friends were in an honors math class in 6th grade, and she wanted to join them. Alicia's teacher agreed she was ready for more challenging content in math and connected her to the assistant principal, Derek, to see what he could do.

Derek was tasked by the school principal with making the schoolwide schedule, which is commonly referred to as a master schedule. Consequently, Derek's decisions of when to schedule which courses based on teacher class load and student class needs would determine students' class schedules. Yet when Derek looked at his options for moving Alicia's schedule, he balked at making the change as Alicia still needed some language support and could not pass the ELD test. He worried that Alicia could not keep up with the access to the math content given the language demands of the class. Besides, if he moved Alicia, there would be a domino effect in the school's schedule. He would need to add Alicia to a class that already had the maximum number of students, and adding Alicia to a math class at another time in her schedule was not an option as she needed the sheltered courses in other subjects. How could Derek meet the needs of Alicia within the confines of the current school schedule?

Derek was frustrated by the situation, because he wanted to place Alicia in the right math class but felt constrained by her ELD needs and her enrollment with other existing sheltered courses. He noticed the pattern of students like Alicia being scheduled within the same courses within a school schedule. Derek saw year after year that he had trouble placing students labeled as English Learners in courses mixed with students who were not multilingual. For example, even if Derek wanted to mix students like Alicia with other students during PE, the schedule limited his options of how to design the school schedule.

Alicia and Derek's experiences are what many students and leaders in schools confront when working to improve students' access not only to math courses but more generally to the content across courses and school programming also. School schedules are one of the ways schools both liberate or segregate students' learning experiences. A school schedule may *show up* in a school leader's operational managerial work that a school leader may not believe to be the most important part of schooling—and yet without intentional design, school schedules may contribute to within school segregation. This chapter examines the other main *actor* in this story: the school schedule. It starts by giving an overview of what is involved in school scheduling: how it works, who is involved, and what the puzzle looks like in action. Then the chapter examines what is known from research when school scheduling goes wrong. When do school schedules exclude students from accessing math or other content, and what are other pitfalls to look for when working to detrack your school schedule? Finally, the chapter explores design principles for schoolwide scheduling that can help liberate students, teachers, and leaders from the tyranny of the schedule to support a detracked system of math courses.

Before discussing school scheduling, it is important to emphasize that changing a school schedule alone will not change teachers' instruction or student learning. The chapters so far have described curriculum, professional development, and coaching, which are all practices needed to change teaching and learning. Elmore (1995) suggested that changing schedules is a form of structural change in school reform and that changes in structures do not necessarily result in changes to teachers' practice. However, changing school scheduling is a little bit more complicated than a structural change. Therefore, the chapters describe scheduling as either limiting or expanding access to courses for teachers and students and plays a role in the ecosystem of factors around detracking that needs to be attended to.

·········································································································

## QUESTIONS TO CONSIDER WHILE READING THIS CHAPTER

This chapter explores a few essential questions, organized to provide an understanding of the larger context related to tracking:

**The basics on schoolwide scheduling:** What is involved in school scheduling? Who is involved in developing the schedule? What are the design principles guiding a school's schedule design process?

**The impact on school scheduling for students:** What does a school schedule look like in action for a student? If you want to make a change to a student's schedule, what does that entail? What are the principles guiding this process?

**How to make equity central while detracking a school schedule:** How do you center equity when designing a school schedule? How can you work to achieve heterogeneous classrooms when creating a school schedule? What are common pitfalls or barriers to supporting equity and heterogeneity in schoolwide scheduling?

## THE IDEAS, IMPACT, AND INNOVATIONS FOR DETRACKING COURSES WHEN DESIGNING SCHEDULES

The ideas in this chapter will help teachers, school and district leaders, community members as well as families or caregivers and students learn the basics of school scheduling and explore innovations in scheduling related to detracking courses. The ideas in this chapter define school scheduling, explain the process of making a school schedule, and center school scheduling on equity and access.

Also, a lot of the research and stories from the field explored take place in secondary school. This is not to say school scheduling in elementary school does not influence tracking, and this chapter does discuss at some points scheduling and detracking in elementary schools. However, the structural influences of tracking in secondary schools are more documented and pronounced given the puzzle of organizing students and teachers taking and teaching multiple classes throughout the week. So next, the chapter starts by defining school scheduling.

## School Scheduling Is a Process of Sorting Students and Teachers

Scheduling is a dynamic part of how schooling is organized. When school leaders create a course schedule in a secondary school, they are creating a plan for how teachers and students will sort into various courses. In essence, the plan is a procedure—a student shows up to a certain classroom with a specific teacher with other students at a certain time focused on a class curriculum and stays in the class for a set amount of time (e.g., quarter, semester, or a year) over several days within a week. Other classes happen simultaneously. Often, bells ring at the start of class and end of class to indicate it is time to move to the next activity, whether that be another class, lunch, activities, or the end of school. Classes may last for a more standard time (40–60 minutes) or classes are organized in longer blocks of 80 to 110 minutes. Classes meet sometimes daily in the standard time, block times might be every other day, or a combination of both kinds of times, meeting two to three times a week. Scheduling of classes changes either on a semester (approximately 16 weeks), trimester (10–12 weeks), a quarter (usually 10 weeks), or over the whole school year.

Another way to think about scheduling is what Domina and colleagues (2019) call "school-level sorting practices." This research team set out to develop measures to understand differences in these sorting practices within schools. They tested five measures of sorting students into courses or *tracks* of courses: (1) differentiation in curriculum taught within a class, (2) the level of heterogeneity in student skill levels within courses, (3) the rate of student enrollment in courses teaching *advanced* skills or skills beyond the stated grade-level standards-based requirements, (4) the extent to which students move between more or less advanced courses or tracks over time, and (5) the relationship between track assignments across subjects. Domina and colleagues' measures of these sorting practices imply that class scheduling involves sorting students and teachers based on

- the design of the class content (curriculum)
- the *level* of the class content above, at, or below grade-level standards
- students' exposure within courses to students with different skill levels

- students' movement between courses at different levels

- students' movement between class level across courses on different subjects

It is important to acknowledge that Domina and colleagues' ideas in and of themselves may be incomplete. For example, in the San Francisco Unified School District (SFUSD), there are the goals of keeping cohorts of students in families (or houses, like freshmen house, sophomore house, etc.) and academies (usually in 11th or 12th grade) that heavily influence scheduling and do not fit neatly into the idea described by Domina and colleagues. Also, the ideas only capture characteristics that may be easily measured by researchers and do not capture system-level sorting like requirements that students take certain courses based in an Individualized Education Plan (IEP) or other specialized supports for students.

Another important variable influencing a school schedule is how student and teacher characteristics and preferences may influence the sorting of students and teachers into classrooms. For example, Domina and colleagues point out that they do not explore students' differences in demographics and identities like race, class, gender, and overall mindsets toward a subject like math, science, English, and so on that might be overlooked by their research. It is important to emphasize that a student's relationship with a teacher, including having a previous class with a teacher or racial match with a teacher, could influence student preferences and consequently sorting. Similarly, differences in teachers' own demographics and mindsets toward a subject or class may influence their assignment to teach a class. For example, some teachers may prefer teaching more advanced courses with content taught to students that exceeds their grade-level standards. Teachers may believe that the students in advanced courses have certain characteristics that are different from students in courses designed to cover grade-level or below-grade-level content.

It is important to note that most of these variables influencing schoolwide scheduling apply to secondary school contexts, but there are some cases where student and teacher characteristics and preferences influence elementary school scheduling as well. For example, a school schedule has to be coordinated at the elementary level for special courses like PE, music, and art. Also, there is certain programming in elementary schools that needs to be scheduled like support within the classroom from a teachers' aide, instructional coaching, or programming that pulls students out of courses for English Language Development (ELD) courses or specialized courses for students with IEPs. This may lead to classroom-level sorting based on language status, reading level, or other sorting markers for students.

# LESSONS LEARNED

School-level sorting practices supported by scheduling can provide supportive conditions for student learning and may create barriers at the individual, classroom, and school systems levels.

## School Schedules Are Created by Personnel in Different Roles and Using Various Processes

There are no specific rules that dictate who makes a schedule and what process must be used. Some schools delegate the task to an administrator like the principal or an assistant principal. Other schools assign the task to a committee involving community members with different roles, like principals, teachers, counselors, other staff, and even students. Some schools centralize the process into having one or two individuals develop the schedule and present it to the staff and students for review and approval. Other schools democratize the process by having a committee of people design the schedule and having staff and students vote to approve the schedule.

One of the more democratized approaches to developing a schedule is called *site-based scheduling*. It is among a few approaches to designing a school schedule that was more widely documented during the 1990s, and it attempts to involve multiple community members in gathering a wide range of input while developing various iterations of the schedule. One article by Craig (1995) describes the approach of site-based scheduling in a step-by-step process. Craig argues that site-based scheduling relies on the premise that school schedules should be created by the people most impacted by the scheduling—the staff and teachers. There is one important modification to make to Craig's description of site-based scheduling—the committee membership and community input could include students as one group in a school most impacted by a schedule. It is important to caution that it can be politically tricky to involve a committee that has differing opinions, and the process of site-based scheduling can take time and resources.

The steps Craig outlines include

1. *Forming a schedule committee that is voluntary in nature.* The members need to be committed to the process for developing a schedule, but they don't need to be representative of every single viewpoint.

2. *Understanding the goals the committee hopes to achieve when designing the schedule.* To determine those goals, the committee conducts a needs assessment through a survey of all staff and students. The data will be used to set goals. Craig reminds readers the overall goal of scheduling is to have a "flexible schedule that meets as many needs as possible" (p. 17) so the data collected in the needs assessment will help reveal the different needs from staff and students influenced by the schedule.

3. *Prioritizing goals for scheduling based on the data from the needs assessment and other evidence collected by the committee.* The committee could develop a draft set of goals based on the data from the needs assessment. Then they could compare these draft goals with other evidence they collect by attending conferences or talking to other schools. Craig suggests presenting the data to their staff or other groups for further input of the draft goals.

4. *Finalizing the goals based on the other evidence collected and developing a draft schedule.* The committee may adjust their goals to some of the realities they come up against with scheduling, such as certain space constraints (number of classrooms) or staffing constraints (number of teachers teaching certain subjects in a given year). The committee members may make a spreadsheet outlining class and teacher assignments based on time and day.

5. *Presenting the draft schedule to staff, teachers, and students, and adjusting based on their input and feedback.*

## LESSON LEARNED

Schedules need to be as flexible as possible to meet the needs of a school community, and the process used to develop the schedule can influence the effectiveness of the schedule.

## Rules About Access Influence Scheduling and Opportunities

In addition to the practice of site-based scheduling, a lot more is known about the impact school schedules have on students' access to courses. For example, to understand the consequential nature of leaders' decisions on the courses students take, let's look at how these decisions shape the access of a growing subgroup of students across U.S. schools—multilingual students who are bureaucratically labeled English Learners. Then we can explore ideas for how to create schedules that provide equitable access to grade-level and advanced class content and access to a diversity of peers within classes.

Estrada (2014) studied multilingual students' access to courses across four middle schools demonstrating how school leaders' decisions about students' class placement influence multilingual learners' opportunities to take certain courses in their scheduling. As seen in Figures 8.1 and 8.2, Estrada describes two schools' curricular streams for multilingual learner students in graphic form. Curricular streams outline students' access to core content at their grade level, ELD instruction, and the balance between remediation and acceleration (e.g., support classes), as well as isolation and integration. Figures 8.1 and 8.2 show how these two middle schools sort multilingual learner students into courses disaggregated by their English language proficiency on an assessment, their years in the U.S. school system, and their achievement on a state standardized test in English Language Arts. The two figures also show the complexity of creating equitable class pathways for multilingual learners. For example, Figure 8.1 shows more limited access to core content because of ELD courses and multiple remedial courses supplanting core courses. Students across different levels of English proficiency have different levels of access based on the courses they take. For example, students labeled with beginning proficiency taking most classes labeled ELD. Students labeled as early advanced proficiency take Specially Designed Academic Instruction in English (SDAIE) courses and few mainstream courses where multilingual learners are with their non-multilingual peers.

## FIGURE 8.1 ONE SCHOOL'S APPROACH TO ASSIGNING STUDENTS LABELED AS "ENGLISH LEARNERS" TO COURSES

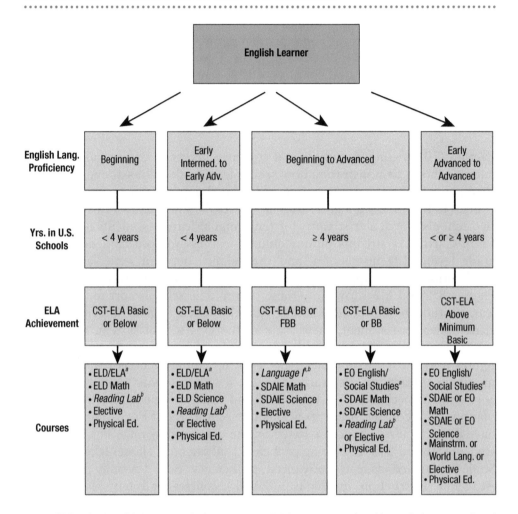

Cove Middle School English learner *Curricular Streams*. English learners were placed in *Curricular Streams* based on English language proficiency levels as assessed by the California English Language Development Test, years enrolled in U.S. schools, and performance on the California Standards Test-English Language Arts (CST-ELA). CST-ELA performance levels are Far Below Basic (FBB), approximately 3 years below grade level; Below Basic (BB), approximately 2 years below grade level; Basic, approximately 1 year below grade level; Proficient, grade level; and Advanced, above grade level. ELD denotes English Language Development; ELA denotes English language arts; EO denotes English only. SDAIE denotes specially designed academic instruction in English. Twenty-two percent of ELs enrolled in the two ELD/ELA streams; 78% enrolled in the other three streams. [a]These courses were two-period blocks. [b]These courses were reading interventions.

Source: Estrada, P. (2014). English learner curricular streams in four middle schools: Triage in the trenches. *Urban Review, 46*, 535–573. https://doi.org/10.1007/s11256-014–0276-7 Copyright by Peggy Estrada. Reprinted with permission.

In contrast, Figure 8.2 shows a school schedule that attempts to provide access to core content and integration into mainstream math and social study courses. It provides fewer barriers to accessing core content in English language arts (ELA) courses by using English proficiency as the only criterion for accessing mainstream courses.

**FIGURE 8.2   ONE SCHOOL'S APPROACH TO ASSIGNING STUDENTS LABELED "ENGLISH LEARNERS" TO COURSES**

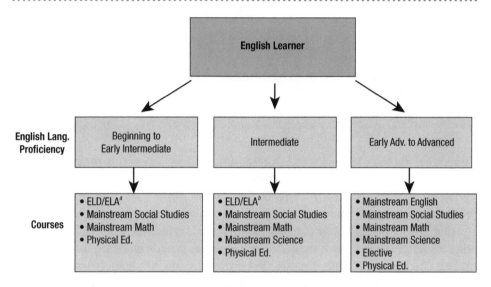

Interior Middle School English learner *Curricular Streams*. English learners were placed in *Curricular Streams* based on English language proficiency levels as assessed by the California English Language Development Test (CELDT). ELD denotes English Language Development; ELA denotes English language arts. Fifty-four percent of Els enrolled in the two ELD/ELA streams; 46% enrolled in the mainstream *Curricular Stream*. [a]These courses were a three-period block. [b]These courses were a two-period block.

Source: Estrada, P. (2014). English learner curricular streams in four middle schools: Triage in the trenches. *Urban Review, 46*, 535–573. https://doi.org/10.1007/s11256-014–0276-7 Copyright by Peggy Estrada. Reprinted with permission.

Estrada points out that school-level decision making played an important role in influencing these curricular streams and consequently access to the courses. For example, Estrada described how one school's leaders and teachers were not happy with the curricular isolation of multilingual learner students and, consequently, were more innovative with schoolwide scheduling. As seen in Figure 8.2, the teachers allowed multilingual learner students at the Beginning and Early Intermediate to be integrated with non-multilingual students in mainstream courses like math and social studies. This allowed multilingual learner students in the beginning of their development of English language proficiency to be in classes with other English proficient students. Contrast this with students in their beginning levels of English proficiency at the school in Figure 8.1 who were not allowed access to any mainstream courses. In summary, Estrada's study demonstrates that school-level decisions about class schedules for multilingual learners influence students' access to core content and integration with non-multilingual peers in their courses and those decisions vary across schools.

## LESSON LEARNED

**School leaders' decisions may perpetuate, reduce, and/or eliminate tracking in a schoolwide schedule.**

As seen in findings from Estrada's research, leaders' decisions may intentionally influence students' access to courses based on achievement, English proficiency, and other exclusion criteria. However, sometimes leaders' decisions produce unintended consequences in a school schedule thereby unknowingly limiting students' access to courses. School leaders can look at their data and other research to see how their decisions about scheduling may have limited students' access to certain courses. For example, SFUSD leaders saw research by Estrada (2014), Thompson (2017), and Umansky (2016) examining multiple school districts and suggesting that multilingual students labeled as English learners were inadvertently being tracked into courses with each other because of constraints in their required courses. Umansky (2016) refers to this as "exclusionary tracking" where schools systematically exclude students labeled as English learners from taking certain courses because they are deemed to have lower levels of English proficiency. As seen in the opening vignette, Alicia experienced exclusionary tracking when she was designated as an English Learner and consequently assigned to courses in her schedule based on her English proficiency rather than readiness for core content. These school leaders' decisions about Alicia caused her to be placed in a class schedule that limited her access to grade-level and more advanced core courses and peers not labeled as English Learner.

> *School leaders can look at their data and other research to see how their decisions about scheduling may have limited students' access to certain courses.*

## LESSON LEARNED

School leaders and other community members designing school schedules may unintentionally be tracking students into certain courses based on the constraints of the schedule.

When SFUSD saw this evidence from the research and in their own data, especially the constraints put on students labeled as English Learners, they decided to redesign the schoolwide schedules and access to courses in their middle schools. SFUSD called their middle school redesign *Initiate Wonder* and made block scheduling a centerpiece of the reform, along with universal learning experiences, and more opportunities for professional learning for teachers schoolwide. These changes all hinged on SFUSD leaders reorganizing their school schedules around an equity-centered approach of creating schedules with longer blocks and trimesters so that English Learners and students with learning challenges, students historically underserved by traditional school scheduling, could have more access to electives in their schedules and integration with other peers.

> In the proposed redesign, students will take eight classes per week in a block schedule, which means that instead of 50-minute periods, classes meet longer on alternate days. . . . [I]nstead of only a limited number of English proficient students and students with disabilities participating in what are called "acceleration" courses, all students will have acceleration opportunities that are based on their needs and interests. (San Francisco Unified School District, 2022)

SFUSD leaders say a change in their schoolwide schedule and bell schedule within each middle school is a way to include equitable access to electives and acceleration classes, deeper learning structures like block scheduling, and adult professional learning through aligned early release days across schools.

## Creating Equitable School Schedules by Centering Students

The case of SFUSD's changes to their middle school schedules and access to courses provide one potential roadmap for how to center equity in schoolwide scheduling. In essence, SFUSD leaders were using two design principles to center equity in their schoolwide scheduling: (1) longer class periods with block scheduling and (2) universal learning opportunities—providing core courses, electives, and acceleration classes that all students could access. One might also argue that Craig's (1995) earlier description of site-based scheduling is another form of centering equity during the scheduling process by putting people who are the ones most impacted by the schedule in charge of designing the schedule. However, site-based scheduling may also not center equity by not including voices of students and teachers who are historically underserved by traditional scheduling practices. Here are some promising design principles for creating equitable schoolwide schedules based on some of the evidence about school scheduling:

- *Center the voice of students historically underserved by school systems when designing school schedules:* As mentioned earlier, it is helpful to have a committee of people from different roles throughout the school involved in designing the school schedule. A site-based approach to scheduling promotes the perspective of the end-user of the schedule throughout the design process. A school scheduling committee can support equity goals when they include students and staff who are historically underserved by school systems. This would allow students and teachers, whose perspectives have not been privileged, to weigh in on key decisions about the school schedule. Some schools have been known to have staff *shadow* or follow students through their courses for one day so they can see the school schedule in action from a student's perspective as a means to center the student's perspective.

- *Apply universal design principles:* The concept of universal design stems from building more accessibility into people's everyday lives through their lived interactions with different human-made designs. Architects used universal design to change the way they design doorknobs, moving from a twisting knob to a lever to make it easier for everyone to open a door for a teacher carrying an armload of books or for those who use a prosthetic hand. Scholars like Rose and Meyer (2002) suggested that these principles could be applied when developing instructional materials to make them more accessible to students with certain disabilities. Hitchcock et al.'s (2002) notions of access, participation, and progress or CAST's UDL Guidelines (2018) could be useful when examining the design of a school schedule. For example, to integrate universal design principles into your schoolwide schedules, ask:

  ○ Do all students access grade-level mathematics in their class schedule?

  ○ Does the schedule allow all students to participate in math courses that have clear goals tied to graduation?

  ○ Does the schedule allow all students to progress in their secondary career toward graduation and college and career readiness?

  ○ Does the schedule allow opportunities for all students to experience joy, to be challenged, and to thrive in various settings?

These questions will help school communities examine whether their schedules allow students to access their core content and have the opportunity to progress in school.

- *Explore different arrangements of class time:* When designing schedules, many leaders feel constrained by the number of minutes in a class and the number of courses needed to complete grade-level and graduation requirements. Some school leaders have used creative strategies to break free from these constraints while still meeting class requirements. For example, some leaders have created longer class periods referred to as *blocks* to change their schedules. Other schools have developed zero periods that come before the first period and allow for additional optional sections for students. SFUSD leaders lengthened the school day to make their middle school schedule have more flexibility and access for students. School communities could think creatively about the amount of time classes take and the arrangement of classes in a day.

- *Conduct an audit of students' access to learning opportunities.* Randomly select five to ten students in two to three subgroups and examine their access to courses. Before designing your school schedule, collect data about individual or small subgroups of students with specific characteristics. For example, in Estrada's curriculum streams, she creates subgroups of multilingual learners based on the bureaucratic label of "English Learner," levels of language proficiency, ELA scores, and years in school. Next, try examining the general patterns of those individuals or subgroups of individual students' patterns of taking courses. Finally, compare and contrast the patterns of taking courses. Do the patterns look similar or different? How does that class relate to the core courses for the other students in that grade level? Are the courses students are enrolled in have students with demographics like race or gender that are heterogeneous or homogeneous?

## LESSON LEARNED

If possible, conduct an audit of a school schedule on a semi-annual basis to make sure it is achieving the preset goals of the schedule.

In addition to these design principles, the chapter describes two common pitfalls and barriers to equitable scheduling.

1. *Some schools lead the scheduling process by using data exclusively focused on teacher preferences for teaching certain courses.* In *Catalyzing Change* (The National Council of Teachers of Mathematics, 2018), the authors refer to this as "teacher tracking." Teacher preferences for teaching courses are important because teachers have certain expertise and supporting teacher preferences can help with teacher recruitment and retention. Some school leaders may also find teacher preferences for teaching certain courses a barrier to changing their schedule. School leaders working to reform their school schedules might ask some of their teachers most effective at engaging students and having a high success rate with historically underserved students to look beyond their preferences and grow their practice by teaching courses at certain grade levels or courses with new content to create more access and opportunities for students.

2. *Some schools overlook smaller groups of students' lack of access to certain courses within the design of their schedule by centering their schedule design on the majority of students' needs.* This pitfall is especially true for smaller subgroups of students with special needs— multilingual-learner students, students receiving special accommodations or instruction aligned to their IEP, students who are currently struggling with Ds and Fs, and so on. It is important to make sure that when leaders refer to "each and every student" (from the individual, to the smaller subgroup, to the larger majority), there is a school schedule that ensures all students have access, including those students who are historically underserved by more traditional school schedules.

### Questions to Consider in Your Context

▶ As you examine your school schedule with the aim to make your classrooms more heterogeneous and also provide more access and equity in taking courses, here are some questions you could discuss with your team:

- What are students' experiences when taking courses during the schoolwide schedule? Does the schedule allow them to take the courses they need and want? Do students historically underserved by most school systems have access to core content and the courses they need?

- What are teachers' perspectives on scheduling? What types of courses do they prefer teaching? Do those preferences influence your schoolwide schedule?

- Who is designing your school schedule? Does the design center the end users—students? How does the design of your school schedule center Black, Latinx, multilingual, or other student groups who may not have access to courses because of status quo scheduling practices?

- Is your school schedule helping students achieve their desired goals in math? How do school schedules relate to student engagement and achievement?

## Activity 8: Activities for Exploring Your School's Schedule With Your Teachers

In this section we offer activities for exploring your school's schedule with teachers and leaders using a variety of lenses. One of the groups most impacted by scheduling is teachers. Scheduling influences the pace of teachers teaching, the content they teach, and the students they work with and can either support or constrain their opportunities for professional learning and team planning. In this activity, based on a worksheet developed by the SFUSD Math Complex Instruction team and then adapted by Torres & Woodbury (2021), you will consider ways that your school's schedule can support better opportunities for strong teacher collaboration leading to better learning opportunities for your students. This activity has been used with teacher leaders, department chairs, and site administrators in professional developments and then used as a tool to follow up during site support with these individuals. Depending on where a community is in relation to your math goals, vision, and the amount of collaboration that is currently happening, a leader may focus on one specific area of this worksheet. For instance if course- or grade-level teams have not planned much previously, saving the ideas around peer-reciprocal observations and rotating teachers across course or grade levels might be something to focus on in future years after collaboration is stronger.

*(Continued)*

## Considerations for Scheduling Conversations

Math Goal: To create classroom experiences that support equitable participation while learning rigorous grade-level math.

Consider ways that your school's schedule can support learning opportunities for your math team. (Teams to consider: course teams, cross-course teams, math and neurodiverse collaboration, etc.)

**Reflection:**

- What are the current <u>strengths and areas for growth</u> in your math team (as a team, students, etc.) as related to the district vision for math instruction and site learning communities?

- What are your math team's <u>goals</u>, and <u>how does the team want to learn together</u>?

- What are the parameters of your department's schedule (# of courses, # of sections, # of FTEs, etc.)?

- What is the process at your site? Who are the players involved in creating the school schedule and setting the math schedule?

- What courses are available to students? How are decisions made about who has access to each course? Which students are you serving or not serving in these decisions?

- Is it the math courses, the scheduling of the courses, or other teachers' supports that promote teacher growth, planning, and teaching as well as student access, equity, and achievement goals?

## Common Planning Time

*What time is available (or can be made available) to plan with course teams?*

| Questions to Consider | Notes |
|---|---|
| • When will course- or grade-level teams plan together? <br><br> • When can common prep periods be aligned? Which teams would benefit from having a CPT (common planning time) or a common prep period? Why? <br><br> • When can teachers plan with co-teachers? | |

| Course Teams | |
|---|---|
| Which teachers are teaching which courses? | |
| Questions to Consider | Notes |
| • Do you have a strong Math 6/Math 1/ Algebra 1 team apprenticing students into your math program?<br><br>• Do you have a variety of teacher experiences on each course team (years of teaching, familiarity with curriculum, experience with multilingual or neurodiverse supports)? Are teachers being tracked (experienced teachers only teaching upper level courses, new teachers teaching siloed courses, etc.)?<br><br>• Is there someone on the different course teams who is responsible for holding the vision and community goals in mind during course- or grade-level conversations?<br><br>• Who will lead course teams? Does their teaching reflect the vision the team is working toward?<br><br>• How are you strengthening your collective understanding of vertical alignment over the years by having teachers rotate through different courses over the years?<br><br>• Might some teacher looping be beneficial for students? (For example, this year's Math 1/Algebra 1 teacher(s) teach Math 2/Geometry next year to support continuity and building on relationships.) | |

*(Continued)*

*(Continued)*

| New Teacher Support | |
|---|---|
| Questions to Consider | Notes |
| • How do you use the schoolwide schedule to support apprenticing new teachers? | |
| • Does every incoming teacher at your school have a partner to plan with? | |
| • If you don't have enough classrooms for every teacher, who will travel? How can you distribute this and similar challenges across your team to support the success and retention of new teachers? | |
| • Are there members of your team who might grow through mentoring a new teacher or having a student teacher? How can your team collectively support new teachers and student teachers? | |
| • **Hiring:** How do you clearly articulate your department values and vision so you can communicate it to those who want to be hired? | |
| • **Hiring:** How can you really get to know a candidate, including their beliefs about collaboration, and beliefs about students? What questions can you ask during an interview that are relevant to your team's vision of math teaching and learning? | |

| Opportunities for Peer Observations | |
|---|---|
| Questions to Consider | Notes |
| • Which teaching teams or teachers would benefit from having opportunities to do peer observations? | |
| • How can the schoolwide schedule facilitate peer observations? | |

Source: SFUSD Math Complex Instruction Program (2019).

# CHAPTER 9

·······························

# CONSIDERING RESEARCH THROUGHOUT YOUR MATH DETRACKING JOURNEY

In many schools and districts, teachers are not only teaching mathematics in their classrooms, they are also the ones designing and leading the professional development for other teachers across schools. In San Francisco, Ho Nguyen, one of the authors of this book, was one of the teachers on special assignment hired to organize and execute the professional development for high school math teachers. At the time Nguyen started in this position in 2008, there were no central professional learning opportunities for high school teachers. After visiting many high school math classrooms, Nguyen observed that there were a significant number of math teachers who were trying group-work aimed at supporting more student discourse during instruction, but the teachers were struggling to get all students engaged.

Nguyen also noticed that there were a few math teachers working to address this challenge by using some Complex Instruction strategies; they wanted more professional development. The Complex Instruction approach stems from research conducted by Cohen and Lotan from Stanford University and in secondary mathematics specifically by the math teachers at "Railside High School" (a pseudonym to protect anonymity) researched by Boaler and Staples (2008). Complex Instruction seeks to close the opportunity gap in classrooms by helping teachers design instruction to address student engagement through mitigating students' status based on race, language, class, or academic skills that play out in heterogeneous classrooms. Nguyen and others read Cohen's book published in 1994, Designing Groupwork: Strategies for the Heterogeneous Classroom, and became focused on bringing the principles of Complex Instruction alive in their classrooms. In some cases, teachers were training on Complex Instruction while earning their teaching certificate and master's degrees at Stanford Teacher Education Program or while working in other school *districts. As Nguyen was developing as a math teacher leader, he attended a workshop at the NCSM where he was reintroduced to principles of Complex Instruction, and he became inspired by the need to address inequities caused by status hierarchies that play out in classes on a daily basis.*

*With this interest in Complex Instruction from enough San Francisco Unified School District (SFUSD) math teachers, Nguyen set out to find a way to offer professional development in Complex Instruction to the math teachers open to this work, most of whom were high school teachers in which he was charged to support. He knew he could not execute this work on his own and relied heavily on researchers and practitioners from many different organizations. Here were some of the partners he brought in and their roles at the time they began working with SFUSD teachers: Rachel Lotan, the director of the Stanford Teacher Education Preparation Program and the Complex Instruction program at Stanford University; Lisa Jilk, a professor in the College of Education at the University of Washington and professional developer; Karen O'Connell, a professional developer; Evra Baldinger, a teacher in SFUSD; Carlos Cabana, a teacher at San Lorenzo High School; Estelle Woodbury, a math specialist in Oakland Unified; Kristina Dance, a professional developer; and Nicole Louie, a graduate researcher at UC Berkeley. Nguyen worked with these partners to offer a professional development program for teachers that included a course on designing groupwork, in-class coaching support, collaborative team support, video clubs, and teacher leadership support working in Complex Instruction (Jilk & O'Connell, 2014).*

*As this training proliferated, more and more math teachers wanted professional development on Complex Instruction principles and strategies. Teachers saw how implementing Complex Instruction increased each and every student's engagement in their math classrooms and allowed them as teachers to remove barriers to math content for students who had traditionally not thought of themselves as* math *people.*

*Nguyen also relied on Cohen's book,* Designing Groupwork, *as an important reference for supporting teacher learning. He bought copies of the book for all the teachers attending the professional development series. Nguyen was known for bringing his copy of Cohen's book (which was well penned and earmarked) to reference in meetings and coaching with teachers. In essence, the focus on Complex Instruction in math instruction became an important bedrock for how SFUSD leaders and teachers integrated research into their day-to-day practices while working in detracked heterogeneous classrooms. At one point, Complex Instruction was named in SFUSD's strategic plan.*

## QUESTIONS TO CONSIDER WHILE READING THIS CHAPTER

This chapter explores a few essential questions to provide an understanding of the role of research in the process of developing policies and practices aimed at detracking math courses.

- **Rationale for using research in policy:** Why is it helpful to draw from research, conduct research, and/or partner with researchers in that endeavor?

- **Integrating research in policy and practice decisions:** What does it mean to use research in policy and practice decisions? How is using research in decision making different from monitoring data and collecting evidence to inform decisions? Are there a subset of research studies that lay a foundation for your detracking policy? How is research used when communicating internally or to the public about your policy change?

- **Working with partners and in partnerships to support research use:** Are there certain partners like researchers, third-party organization leaders, or internal staff who are adding to this body of research? Do you have any ongoing partnerships with organizations supporting your school or district to conduct research on your policy and practices used while detracking your math courses?

## THE ROLE OF RESEARCH, DATA, AND EVIDENCE IN DECISIONS

While reading this chapter, it is important to keep in mind the difference between the terms *research, data,* and *evidence*. **Research** is the systematic examination of a certain phenomenon or element in hopes of advancing what is known among the research community or what academics call *theories*. Research involves using an approach called the *scientific method*, where a researcher forms a hypothesis and then checks that hypothesis by performing an experiment or observing a phenomenon in strategic intervals. The community of researchers respects the research findings once other researchers review the research methods and findings without knowing who wrote the research (a *blind* review) to make sure the methods and findings are sound.

This contrasts with **data** and **evidence**. Data refers to information collected to support analysis. While data is used in research, all data do not equate to research. For example, data could be collected in a very unsystematic way for purposes other than research. Imagine collecting data on students' perspectives on math courses to see if they enjoyed their courses, but only asking students who attend first-period math courses, which would limit the conclusions drawn from the data about *all* math students in the school. **Evidence** is a much broader term than data or research. The term *evidence* means information collected to inform a conclusion. As seen in Figure 9.1, a body of evidence could include data, but it could also include research, personal experiences, or even polling of people's perspectives. You might conduct empathy interviews or have a town hall discussion with families and students.

**FIGURE 9.1  EXAMPLE OF THE PIZZA PIE OF EVIDENCE USED IN DECISION MAKING IN EDUCATION**

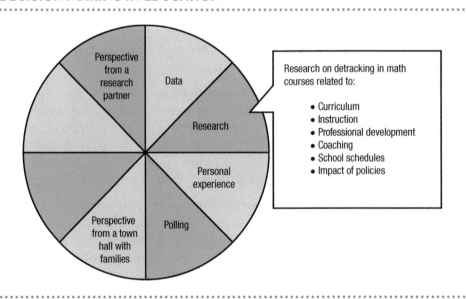

## There Are Different Types of Research

In teachers' and leaders' decision making about detracking, they may use multiple forms of evidence to triangulate with the research findings. Yet why is research essential to include as one type of evidence? Two reasons: Unlike other forms of evidence, research is conducted using systematic methods, like the scientific method (what people might think of as scientific experiments). Researchers have specialized training to help them develop hypotheses they want to test and questions they want to explore, compare their ideas to existing research, and then test their theories by collecting and analyzing data. Also, research that is published in journals and consequently accepted as theory goes through a blind review. This involves researchers critiquing the research without knowing who conducted the research, and editors either accepting the research for publication to journals or rejecting the research based on the blind reviews. While all research has some subjectivity, generally, research published in peer-reviewed journals is accepted as a form of evidence that is less biased than other forms of evidence, such as one person's perception of phenomena or data gathered by people in favor of a practice or policy.

Similarly, there are other systematic ways of conducting research that don't use the scientific method but are often used by practitioners in education. For example, there are traditions of using continuous improvement practices to systematically examine practice using specific theories of change, collection of data using specific measures, and plan, do, study, act inquiry cycles (e.g., Bryk et al., 2015). Also, some educators use an action research approach to examine their practice (e.g., Stringer & Aragon, 2020), which centers research that helps move practical outcomes by collecting evidence that motivates certain changes and behaviors. The traditions of continuous improvement and action research can be important practices for practitioners to collect their own evidence that help them compare to other existing research.

## LESSON LEARNED

It's important for the school and district leaders to collect their own evidence to triangulate with findings from data produced by researchers.

## Research Is One Slice of Evidence

School and district leaders factor in research as one slice of the pizza pie of evidence used in education decision making (modeled in Figure 9.1). This larger body of evidence, which may include things like polling, personal experience, and the perspectives of a variety of community members, could enhance a broader understanding of the design, development, and implementation of detracking of math courses. In one study of a school district deliberating over a policy decision, district leaders used ten different types of *reasons* to explain their thinking. Two of the reasons were data and research. The other reasons included content-specific beliefs, design conditions, guiding principles, policy, among others (Huguet et al., 2021). Of the 1,000+ different reasons used by district leaders, only 3.5% of the reasons were research, whereas community considerations and design conditions were the larger portion of reasons used. In Chapter 10, we go into greater detail about collecting evidence to monitor the policy using continuous improvement practices.

## THE REALITIES OF USING RESEARCH IN POLICY AND PRACTICE DECISIONS

There has been some documentation for how research is used in social sector policy and practice decisions. In Weiss and Bucuvalas's (1980) book titled *Social Science Research and Decision-Making*, they explain the different approaches used by social sector leaders to integrate research into their decision making. These ideas have been further developed and explained by Coburn and Stein (2010) in their book titled *Research and Practice in Education: Building Alliances, Bridging Divides*. As seen in Figure 9.2, Weiss described instrumental use of research, or research used to directly change policy and practice; conceptual use of research used to change the way people think; symbolic use or research evidence that is used to justify an existing policy or practice; and political use or research evidence used to advocate for a new policy or practice.

FIGURE 9.2   FOUR WAYS RESEARCH INFORMS POLICY

| | |
|---|---|
| **Instrumental:** research evidence that becomes the motivating factor for a change in policy and practice | **Conceptual:** research evidence that requires people to change the way they think about a policy or practice |
| **Example:** Research on detracking math courses through a certain grade level shows changes in student achievement that motivates a new detracking policy. | **Example:** Research on tracked math courses may convince people (or dissuade depending on the findings) that detracked math courses is a good practice. |
| **Symbolic:** research evidence that is used to justify an existing policy or practice | **Political:** research evidence that is used to advocate for a new policy or practice |
| **Example:** Leaders at a school or district share research that justifies a policy or practice that is already being implemented. | **Example:** Leaders at a school or district point to research that puts a prospective practice or policy related to detracking in a positive light. |

Source: Adapted from Coburn, C. E., & Stein, M. K. (Eds.). (2010). *Research and practice in education: Building alliances, bridging the divide*. Rowman & Littlefield; Weiss, C., & Bucuvalas, M. (1980). *Social science research and decision-making*. Columbia University Press.

Arguably, this is not an exhaustive list of the approaches to using research in policy and practice decisions. Yet we believe this framework of instrumental, conceptual, political, and symbolic research use will be helpful as you think about the role of using research in your decision making about your detracking policy. Next, let's explore using research in the design, rationale, and evaluation of the policies and practices in a detracked system of math courses.

## USING RESEARCH TO DESIGN NEW POLICIES AND PRACTICES

When considering a policy aimed at detracking math courses, you can use research as one source of evidence to inform the policy design. As suggested in earlier chapters, a detracking policy needs to attend to instruction through curriculum, coaching, professional development, and school schedules. There is research in each of those topics that could also be considered. What research should your school, district, and community leaders pay attention to?

You could use existing research in an instrumental way to inform certain design features in your detracked math policy. As discussed in Chapter 8, you could use research to influence the design of your school schedule keeping heterogeneous classrooms at the center of the design. You could point to the research by Estrada (2014), Thompson (2017), and Umansky (2016) showing some unintentional exclusionary tracking as a motivating factor for redesigning school schedules that ensure multilingual learner students are placed in heterogeneous classrooms.

If you do not feel equipped to access and judge the quality of existing research, you could find a researcher or others steeped in math education research to synthesize the research for you. Some of these research partners will spend time talking to you about their overall knowledge of the research in hopes it supports school, district, and community members' understanding and use of the research. If the researchers are faculty from universities, this engagement with schools and districts may count toward their public service when they apply for tenure. Other research partners enjoy being a part of the policy discourse and make their engagement in the policy development part of their research process. For example, as mentioned in earlier chapters, SFUSD leaders relied on Phil Daro and Harold Asturias from the Strategic Education Research Partnership (SERP) and UC Berkeley as thought partners who supported their development of a position paper outlining the research that backs a detracked math policy. Daro and Asturias helped SFUSD leaders identify and synthesize research from multiple sources within their position paper.

In the opening vignette, we see Nguyen using research by Cohen and Lotan to design new practices. Cohen's book, *Designing Groupwork*, was originally published in the 1980s and is based on years of research and many studies conducted by Cohen and later her colleague Lotan to test the efficacy of their instructional approach, now called Complex Instruction. Given the strength of the approach, Cohen's book is now in its third edition with Lotan as a co-author. Starting in 2009, Nguyen brought in consultants like Lisa Jilk from the University of Washington to lead professional development on Complex Instruction, which became a robust program for interested site teams (Jilk & O'Connell, 2014). Nguyen was able to take the learnings from this program and research to influence secondary math teachers with professional development. Starting in 2008, Nguyen used the principles from

Complex Instruction to design the professional development for all high school math teachers. One of the design features of Complex Instruction—heterogeneous grouping of students—became a design feature that not only influenced the math professional development for SFUSD teachers but also motivated Nguyen and other leaders to advocate for detracked math courses.

Using research in the design of the policy and practices was only the beginning for SFUSD leaders. As you will read next, they went on to use research to support the passage of the math policy and its evaluation and improvement over time.

## USING RESEARCH TO PROVIDE A RATIONALE FOR A NEW POLICY

In addition to the design of the policy, you can also use research to support the decision to adopt a new policy to detrack your math courses. When using research to support a new policy, it will be important to keep three things in mind when selecting what research to reference.

1. *The research needs to make sense for your own community context.* For example, if you are in California or Texas, it might be helpful to cite research conducted in other school districts within the state. Also, if you are in an education setting where math teachers use a prescribed curriculum, it would be important to understand the research conducted in settings where there were also specific types of curriculums used.

2. *The research needs to make sense to the elected officials and community members.* In most cases in education, elected officials on a school board make the ultimate decisions about math course policies. Your research must be presented in a clear and concise manner at public meetings so that elected officials (and leaders and families) understand the complex issue and benefits to your proposed detracked policy. The research and other evidence must persuade the elected officials and the community members and interest groups who voted those officials into office regarding the merits of your proposed policy.

### LESSON LEARNED

It's better to make your use of research clear to all community members by explaining the research evidence in public presentations to the school board and community.

In SFUSD's case, the math leaders used research to support a policy change involving the detracking of math courses. They referenced research, most of which was conducted outside of SFUSD, in their presentation to the board, the teachers and leaders within SFUSD, and community members outside of SFUSD. For example, they proposed their detracking policy to the SFUSD school board commissioners in their 2013 position paper. As referenced in Chapter 3, the position paper describes the policy change and then uses existing research conducted in other settings to explain the reason why it is important to detrack SFUSD's math courses. For example, the position paper references San Francisco State professor Maika Watanabe's 2012 video and book of resources, *"Heterogenius" Classrooms: Detracking Math and Science*, which features research-backed practices that support the successful implementation of detracked math classrooms. In the Appendix of the position paper, the SFUSD leaders provided an excerpt from Watanabe's book titled "Why Should I Care About Detracking? What Is Wrong With Tracking?" The excerpt summarizes research about detracking and tracking, including citations of Jeannie Oakes's research among many others.

## LESSON LEARNED

Some researchers are willing to engage in the public discourse related to a detracking policy. This can be a helpful support when working to communicate the rationale for the policy to the public.

3. *The research needs to make sense to the teachers, families, and students.* When using research to support a policy change, think about the people who are on the frontlines of and most affected by that change. If the research you use to argue for a new policy does not make sense to them, your use of research may not be that effective. One way to support teachers', families', and students' understanding of the research is to have events and communications that share the research in accessible ways. For example, researchers may be willing to come talk directly to teachers, students, and families about their related research findings. In fact, SFUSD hosted an event for families with Stanford University professor Jo Boaler speaking about the importance of SFUSD policy changes in mathematics, citing her own and others' research on the value of heterogeneous classrooms. Also, SFUSD leaders explain research evidence related to detracking by citing summaries of research and policy recommendations

from professional associations like the National Council of Teachers of Mathematics, in its book *Catalyzing Change* (2018), and other research briefs that discuss detracking. Other professional organizations have their own position papers including NCSM (2020) and TODOS (2022), an organization that advocates for equity and excellence in mathematics. The position papers from these professional organization may be more palatable readings for practitioners and often summarize research.

In another example, many teachers trained in Complex Instruction gave families an understanding of research being used to support the detracking policy. These teachers had experienced first-hand the effects of using a research-backed approach—Complex Instruction—in their classrooms. Consequently, these teachers understood the rationale for the wider policy that created math courses with heterogeneous groups of students. Hence, during the deliberation over SFUSD's new math policy in 2014, many math teachers spoke up to the school board during public comment to advocate for detracked math courses.

## PARTNERING WITH RESEARCHERS TO EVALUATE A NEW OR EXISTING POLICY

One way to achieve instrumental use of research while planning and implementing a detracking policy is by partnering with research partners to conduct research about the policy design and implementation and consequently evaluate the impact of the policy. SFUSD leaders had thought partnerships with researchers equated to having a "researcher on speed dial" (Penuel & Gallagher, 2017, p. 1). It is true that, broadly speaking, SFUSD leaders have had a lot of researchers acting as their thought partners to support their research use. However, the mechanism that supported SFUSD's leaders' instrumental use of research has been their partnerships with researchers to evaluate their math policy.

When school, district, and community leaders work on studies with researchers, it is a complicated endeavor. One way school and district leaders work to simplify working with researchers is by developing research-practice partnerships. These partnerships are long term in nature and work to develop relevant research based on the needs of the educators and community members involved, support improvement in the partner schools, districts, or organizations, while also producing generalizable and rigorous research (Farrell et al., 2021). As mentioned previously, SFUSD had a research-practice partnership with SERP that was established in 2007 and helped SFUSD leaders develop relationships with some local and national leaders who could act as thought partners as the district leaders develop their policy. SERP was also involved in connecting SFUSD leaders to relationships with researchers who could study SFUSD's detracking policy once the policy was being implemented, including researchers from SF State University, SRI, Stanford University, the University of Chicago, UCLA, Northwestern University, and WestEd.

For example, SERP connected SFUSD leaders to Gudelia Lopez and Marty Gartzman (2016) at the University of Chicago who examined how schools and teachers are experiencing the implementation of the SFUSD's math policy. They interviewed twenty individuals in 2016, 2 years after the implementation of the policy. The interviews found themes like reported increases in student discourse about mathematics and stronger student engagement in rich problem solving. Teachers were more often using the new curriculum and student-centered strategies and were more willing to collaborate. Participants cited less concern from parents about the math course sequence. All interviewees reemphasized the vision for an equity-centered math program. The challenges reported by interviewees included teachers needing more professional development to implement heterogeneous classrooms and principals wanting to improve their understanding of the new policy to improve their support for the practices required for implementation. SFUSD leaders used this report to fine tune their support for teacher and leader professional development and coaching to improve implementation of the new policy.

Two other research-practice partnerships have helped SFUSD leaders evaluate their detracking policy. For example, SFUSD was part of a network of school districts in California called *Math in Common*, which was facilitated by California Education Partners and partnered with a nonprofit research firm, WestEd, to examine the effects of their innovations when implementing the CCSS. WestEd (2022) published a report on the findings from across school districts in the network. The WestEd researchers met with each district, including SFUSD, to review the findings from the report and help SFUSD leaders use the findings to improve the policy implementation. For example, the WestEd findings demonstrated that SFUSD's development of a curriculum was essential but not sufficient for improving instruction, so SFUSD made sure to strengthen its coaching for teachers. SFUSD also worked with the Stanford-SFUSD Partnership, also managed by California Education Partners, to have a third party examine the findings from students' course trajectories and access to math courses given the changes in math policy. The SFUSD leaders worked with economics professor Tom Dee at Stanford University's Graduate School of Education, with the intent of using the research findings to examine the slight adaptations to the policy like providing a variety of ways students could accelerate like doubling up two courses, taking a compression course, or taking accelerated summer geometry between 9th and 10th grade.

These research-practice partnerships with SFUSD did not develop overnight. Many of these partnerships were in existence well before the development and passage of the policy that detracked math courses. The partnership with SERP and SFUSD started in 2007, and the partnership with Stanford University and SFUSD started in 2009. In this case, the research-practice partnerships were resources the district leaders could use to help them evaluate their policy.

## LESSON LEARNED

It takes time and resources from school or district leaders to work in partnership with a researcher or a research team, and these partnerships provide important research to inform decision making.

As your community thinks about using research to design, support, or evaluate a policy to detrack math courses, remember a few key ideas.

1. Research is only one type of evidence that is important when developing a policy.

2. Research use can come in many forms (instrumental, conceptual, political, symbolic) and may also take place at different stages of the detracking efforts (designing, rationalizing, evaluating).

3. School, district, and community members can develop research-practice partnerships with researchers to support ongoing research and development when implementing policy and practice changes.

### Questions to Consider in Your Context

▶ As you think about the ideas and lessons in this chapter, here are some questions to investigate in your own context. These questions may be used for hosting a conversation among your design team or a larger group depending on the phase you are in while detracking your math courses.

- Are there researchers in your local community whom you have worked with before? Do you consider those relationships long-term or trustworthy?

- Do you have prior studies conducted in your school or district about your math practices or policies that could relate to your detracking efforts?

- Are there two to three research studies you believe justify the policy change toward detracking?

- Do you plan on conducting any research examining the changes involved in your detracking policy? If so, what are your hypotheses about what you might find from these studies?

- Once you have findings from these studies, how will you share the research with the larger community?

# Activity 9: Summarizing Research Used to Support Your Policy

▶ Once school, district, and community leaders have decided which research they plan to reference in their support for their detracking policy, it is important to explain this research to community groups. We recommend communicating the research in accessible ways and in multiple formats if the research is one of the main forms of evidence to justify the detracking of math courses. Most research is not written in an accessible format; the research is usually summarized in technical articles explaining not just the findings but also the methods in academic ease. These articles are also published in journals that have paywalls that require subscriptions or payment to access.

Leaders can make the research more accessible using a few different strategies.

1.  They may summarize the key findings in straightforward language in short one-to-two-page briefs.

2.  They can share shorter abstracts summarizing the key findings in their newsletters or other communication channels to families (e.g., superintendent's videos or principal news briefs).

3.  They can host events where they share the briefs or abstracts and make sense of the findings with other community members like at Parent Teacher Association meetings, School Site Council meetings, or other community forums like webinars or social events.

As an activity, try taking the research you are using to support your detracking policy, and write it up in three to five sentences or one to two paragraphs.

_____

_____

_____

_____

_____

_____

_____

# CHAPTER 10

......................................

# MONITORING AND MAINTAINING A DETRACKED MATH POLICY

It was February of 2016. Teachers from a large elementary school in San Francisco cohosted a family mathematics night with team members from the San Francisco Unified School District (SFUSD) Math Department. The principal of the hosting school, Ying Kasner, known as Mimi, and Assistant Principal Son-hui Wong, known as Sonny, welcomed families and introduced the teachers. To have parents experience the mathematics the way students do, the teachers decided to use a card game to lead a standards trace across the grade levels of the Common Core State Standards for Mathematics, which were still relatively new to many people.

While the team had anticipated that as many as 60 family members would attend, more than 250 people showed up, including some from a neighboring school. This school had a large community of Asian families, including native speakers of both Cantonese and Mandarin. The event included both formal district-sponsored translators and the informal translation of family members and neighbors and educators translating for one another.

Across the grades, the teachers modeled grade-appropriate versions of the SFUSD Math game Clash—featured by other names in many elementary curricula—a game with a deck of cards usually numbered 0 to 9 for the younger grades, with different variations of numbers on the cards as the complexity of the operations and comparisons build. This game is designed to build numeracy, or fluency and flexibility with numbers. The classroom teachers and the math content specialists supporting the development of this evening knew that this particular game could model how the standards build across the grades and also how manipulatives support students to make sense of their thinking and show their thinking to others.

The kindergarten and 1st-grade teachers modeled a version of the Single Addition Clash. With the exception of the many elementary students who came to school that night with their families, most of the people in the room *did not practice or develop automaticity with arithmetic this way when they were children. Along with the cards, the kindergarten and 1st-grade teachers had bins of linker cubes so that the family members learning in the role of 5 and 6 year olds could model their addition as they played, for example modeling 3 + 5 as three blue linker cubes hooked to five yellow linker cubes. Figure 10.1 shows the directions to the Single Addition version of the game, which the teachers had printed in large poster format.*

**FIGURE 10.1**

· · · · · · · · · · · · · · · · · · · · · · · · · · · · · · · · · · · · · · · · · · · · · · · · · · · · · · · · · · · · · · · · · · · · · · · · · · · ·

### Directions

1. Shuffle the cards. Place the deck number-side down on the playing surface.

2. Each player turns over two cards and then finds the sum of the numbers.

3. The player with the highest (or lowest) sum wins the round and takes all the cards.

4. In case of a tie for the highest (or lowest) sum, each tied player turns over two more cards and calls out the sum of the two cards. The player with the highest (or lowest) sum then takes all the cards from both plays.

5. The game ends when not enough cards are left for each player to have another turn.

6. The player who has the most cards wins.

—SFUSD Math Core Curriculum

Rules for Clash and Variations

· · · · · · · · · · · · · · · · · · · · · · · · · · · · · · · · · · · · · · · · · · · · · · · · · · · · · · · · · · · · · · · · · · · · · · · · · · · ·

*The kindergarten and 1st-grade students were jumping up and down and giggling as they watched their parents being taught this math game by their own teachers.*

*The 2nd-grade teachers were facilitating a 2nd-grade version of Clash with four cards per turn, teaching two-digit addition, using base-ten blocks to model their addition situations. The students all knew that the game builds in complexity—what their teachers called* rigor—*when a player decides which card they should place in the tens place and which in the ones place. The 3rd-grade teachers were teaching the same game using one-digit multiplication with arrays. The 4th- and 5th-grade teachers were teaching the game using addition of fractions with four cards per turn, using fraction kits.*

*Again, the students all knew the game builds in complexity when a player decides which numbers should be their numerators and which their denominators. In some cases, the 9- and 10-year-old students were coaching their older family members with this idea. With 8 minutes per station, and a few minutes to rotate between stations, every family member in the large auditorium was able to play every grade-level version of the game, often being taught by their own child's teacher and their own children.*

*In this chapter, you will think through*

- *Moves you can make to monitor your policy through collecting and analyzing evidence and showcasing any early indicators of success*

- *Adjustments to your policy implementation when your analyses suggest you are not achieving the outcomes you set out to achieve*

- *Bringing new people into the community in ways that support both the history and rationale behind the policy*

- *Integrating the detracking policy into other initiatives within your setting*

- *Leveraging the regional and national conversations about detracking to help ground your work in a wider context*

## QUESTIONS TO CONSIDER WHILE READING THIS CHAPTER

This chapter explores a few essential questions, organized around the big ideas related to monitoring and maintaining your detracked math policy:

- How will you monitor your policy? How will you show early indicators of success before your first cohorts of students graduate as 12th graders?

- How will you make adjustments to your implementation, both to course correct if there is a need and to adjust to contextual factors like a shifting policy landscape in your district or your state?

- How will you support people new to your policy—its purpose and implementation both—at all levels of the system? How will you integrate your policy into new or existing initiatives so that it doesn't feel like *another thing that is being done*?

- How can you leverage the national conversation about detracking and therefore build on the legitimacy of your work?

## MONITORING THE POLICY DURING THE EARLY STAGES OF IMPLEMENTATION

Public schools are slow to shift. In the case of a policy that impacts incoming 8th graders, for example, outcomes of that policy such as high school course taking, or postsecondary enrollment, may not show up for years down the road. Professional supports such as coaching and a task-based curriculum can allow a detracking policy to take root and make change for students, and that takes years of deep work. In the policy landscape, you will need to be ready to monitor early implementation and also show early wins.

## Sharing Early Successes

Early in your implementation, you will want to share evidence of early wins, both with senior leaders and with families. This can take many forms, such as surveys of teachers and students and sharing anonymized examples of student work that provide evidence of depth of understanding.

One example from SFUSD is that the math leadership team used a math task to help them understand how students were doing with their task-based curriculum in heterogeneous classrooms. SFUSD is one of many districts who work with the Silicon Valley Mathematics Initiative (SVMI; www.svmimac.org); among SVMI's resources are assessments that include Mathematics Assessment Resource Service tasks (MARS tasks), a project of UC Berkeley, Michigan State, and the Shell Centre in Nottingham, England. In this case, the team partnered with a research organization to analyze student work with an 8th-grade linear functions task and compared the SFUSD results with other districts that were working with MARS and had also used the same MARS task with students, as shown in Figure 10.2.

**FIGURE 10.2   STUDENTS' SCORES ON MARS TASK**

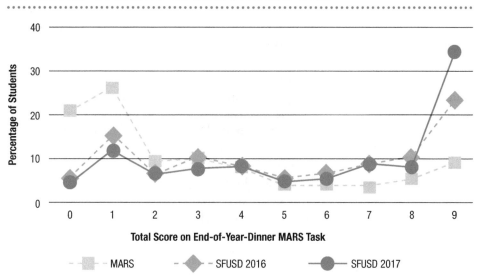

Source: SRI's Final Report: Year 3 SFUSD STEM Learning Initiative Evaluation, *June 2017*. https://www.cs.ucdavis.edu/~matloff/SFUSD/Slides1.pdf

When looking at the graph, the SFUSD team saw that over several years, their 8th-grade students were scoring higher on the rubric when compared to the thousands of students in other SVMI districts that had worked through the same task. The SFUSD Math Team was thrilled to see it reflected back to them that SFUSD 8th graders were able to generalize complex ideas in new situations.

Other early indicators the team looked at were math credits earned by the end of 11th grade, which helped them see if students were on track to graduate and ready for a fourth year of math. They also saw a dramatic reduction of Algebra 1 repeat rates starting with that first cohort of students who were experiencing the new policy and placement practices, which included taking away the barrier of a state standardized test to move on. More of these indicators are shared later in this chapter.

## Monitoring Your Policy

How will you know if schools are implementing a new policy? Early in any policy shift, schools will need clarification and support in shifting practices and sharing resources with families. Disconnects may happen: There may be structural factors within any school that are not obvious to those sitting in central office or on a governing board, and there may be policy features across a whole system that are not obvious to those sitting at a site. Sites and central teams will need to collaborate early and often as they implement a new policy together, and the shifts may take longer than desired.

One example from San Francisco dates to the Fall semester of 2015, or the second year when all SFUSD 6th, 7th, and 8th graders would experience CCSS Math in heterogeneous classrooms. SFUSD STEM executive director Jim Ryan did an audit of course enrollment in all middle schools alongside grades. In that audit, he found one school that was offering two sections of the exact same course at the exact same time; the two sections had an overall difference in average grade point average (GPA) of more than 1.5 points, giving the appearance of a *high* class and a *low* class. With the support of middle school leadership and the site leaders, the analysis went several layers deeper. If students were enrolled in a dual language program, they often travel as a cohort throughout the day so that a single language class can have a cascading effect on enrollment across all content areas. At no point was the intent to shame the school leadership or their math department but rather to better understand the context so that the district might support the school to equitably implement both a detracked math policy and a biliteracy program. Middle school leadership and the site leaders were able to work alongside Ryan to make the necessary corrections and balance the class enrollment. This is one small example of early growing pains that would have had a significant impact on this cohort of students at this school if not addressed with care and respect. You will surely experience uneven implementation as well.

How will you monitor implementation of your policy? What are the kinds of data and stories you can leverage as you move through a time of change? If you want to explore these concepts, there are many places to begin:

1. As you analyze various points of data, find out more about restrictions on sharing student identifying information and be sure to honor any ethical use guidelines.

2. Consider the kinds of assessments already in place in your context. You may be able to use anonymized evidence of student work to help tell your story of change.

3. Consider what points of data will help you understand how your policy is being implemented, especially early in the change process. Most schools and districts have platforms designed to gather and house data such as grades and enrollment. It helps to develop your own fluency with analyzing patterns in the data—both understanding the story of the data itself and how to navigate the platforms where the data is housed.

## LESSON LEARNED

Systems change takes years of deep work. In the policy landscape, you will need to be ready to monitor early implementation and show early wins.

## ADJUSTING YOUR IMPLEMENTATION WHEN NECESSARY

Perhaps you are formally passing a policy to detrack across your district or even your county or state. Perhaps your site-based math department is trying out something new, like detracking one grade level, and visiting each other's classrooms in the process to learn about each other's strengths and practices. No matter the scale of what you are doing, some external factor will certainly exert a force of change. This new factor could be a dramatic localized event like a school shooting or natural disaster that calls on everyone to re-engage in a new way. This new factor could be a shift in the political landscape of your district or state due to elections. This new factor could be a new principal or district administrator who is coming with beliefs about curriculum from work in a different district than your own. This new factor could even be a new statewide framework or shift in legislation about education more broadly. Surely, none of us reading this believed even early in 2020 that a global pandemic would change the way we teach and learn math. All these factors have happened in SFUSD's context; you will need to make adjustments to forces of change throughout your journey as well.

One of the clearest examples of how the SFUSD Mathematics team made adjustments to its implementation is that they expanded the number of acceleration options available to high school students. The district practiced flexibility without losing the core commitment to heterogeneous classrooms, while responding to public pressure from families for more opportunities to accelerate to AP Calculus. The original design had been that all 9th graders would take high school CCSS Algebra 1, with 10th graders taking CCSS Geometry, and 10th-grade students and families having a decision point before 11th grade about whether to enroll in CCSS Algebra 2, or a compression course of CCSS Algebra 2 and Precalculus that would then allow them to enroll in AP Calculus in 12th grade. The team did not include any acceleration options in middle school because it would have meant compressing standards at too early a stage, which would not give students the time needed to experience the depth and rigor of foundational middle grades standards. The district now has three more acceleration pathways open to families at no cost, as described further and as shown in Figure 10.3.

## FIGURE 10.3   CURRENT SFUSD HIGH SCHOOL OPTIONS

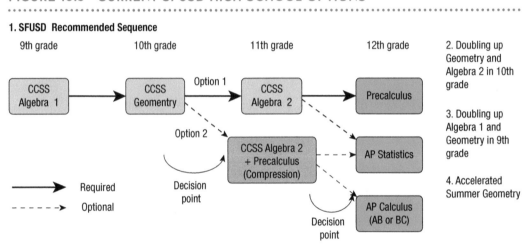

Source: San Francisco Unified School District.

In addition to the compression course

1. students can double up taking two math classes during their freshman year, taking Algebra 1 and Geometry at the same time;

2. students can also double up during their sophomore year, taking Geometry and Algebra 2 at the same time; or

3. students can take an Accelerated Summer Geometry course between 9th and 10th grade, which is effectively another way to double up within one year's high school transcript.

All these acceleration options were kept as heterogeneous as possible; for example, there are no freshmen-only Geometry courses for those who are doubling up in their 9th-grade year. Because enrolling in AP Calculus requires four prerequisite high school math courses, all these acceleration options afford students the opportunity to take Calculus as a senior if it matches their goals.

## The Complexities of High School Math Acceleration Pathways

In SFUSD, while all these four, district-sponsored means of high school acceleration offer options to students and families, each of them has challenges. Both the Algebra 2 / Precalculus compression course and the Accelerated Summer Geometry Course ask students to move quickly through very complex ideas that build and take time. Critiques of each of these courses, both internally and externally, usually point to how hard it is to teach either of these courses at the pace of a student's growing conceptual understanding. One danger is that teachers feel they must skip content to give students the time they need at points in the curriculum; another danger that is the opposite side of this same coin is that teachers *cover* all the standards in the course but feel they must move quickly to procedural fluency at the expense of conceptual understanding. SFUSD tried to mitigate these challenges in the compression course by picking only the necessary prerequisite concepts and skills needed for AP Calculus versus all the traditional Precalculus topics, many of which are not needed as prerequisites for AP Calculus. For the Accelerated Summer Geometry, students have 5.5 hours per day of class time versus the usual 2 hours per day in a summer school course. Both courses present challenges of covering enough material in the given time.

School districts would be disingenuous if they don't offer families ways to accelerate to AP Calculus in high school. We cannot *not* offer acceleration options to AP Calculus in our current system that favors acceleration over deep conceptual understanding, as much as we may lament it and point to decades of research that problematize the Race to Calculus. Some districts have compressed content by pushing high school CCSS Algebra 1 into middle school. The compression problem of moving quickly through complex content persists regardless of where you might compress or accelerate content. To clear a way for students to enroll in high school CCSM Algebra 1 as a middle school 8th-grade student, course developers in these districts have to compress content from two grades into one, or from three grades into two; for example, one can compress Common Core Math 6 and Common Core Math 7, or one could cover a full year of Common Core Math 6 and one semester of Common Core Math 7 while enrolled in 6th grade, and so on. SFUSD leaders considered this even more problematic by tracking at younger ages, by forcing schools and districts to sort into a *fast / slow* or *high / low* track for students who are as young as 10 to 12 years old, before these young learners likely have a sense of their postsecondary goals. This early acceleration would also require more funding, both for more middle-grades math teachers and for an expanded school schedule to accommodate different courses. Similar to the complexities of compression at high school, this can force teachers to rush important middle-grades math content such as linear functions or systems of equations. As a collective of the math education community, we know that tracking directly correlates to racialized outcomes, so tracking younger is even more problematic than offering challenging acceleration options at high school.

Beyond the compression course, the other SFUSD district-sponsored acceleration options at high schools involve taking two high school math courses at the same time: either CCSS Algebra 1 and CCSM Geometry in 9th grade, or CCSS Geometry and CCSS Algebra 2 in 10th grade. Many teachers and schools strongly prefer these doubling-up options, knowing that students can experience a full year of content within the time designed for it. Many families prefer these doubling-up options because students experience less stress and strain with the pacing, though they are of course taking two rigorous math courses at the same time. However, these options also have challenges. Students choosing these options are taking a high school math class in place of another high school elective, such as a world language or an arts class, so they experience fewer opportunities to expand their interests and opportunities. At large high schools that can offer a zero period, students can take two math classes at the same time and still take an elective. However, only large high schools have enough teachers or enough space in their school schedules to be able to accommodate doubling-up options, let alone being able to offer a zero period; therefore, without increased funding, these options are not available to students enrolled in smaller high schools.

One other difficulty about doubling up that may be true in your context is whether you have chosen to arrange the content of your high school courses in a traditional or integrated sequence. There are fewer dependencies between the Algebra 1 and Geometry courses, or Geometry and Algebra 2 courses in a traditional sequence such as SFUSD has, so students in districts with courses arranged in this way can soundly take two courses at the same time. Doubling may not work in districts that have chosen an integrated pathway where content from across the domains of mathematics are taught in the high school courses usually called Integrated 1, Integrated 2, and Integrated 3. For example, some districts are innovating at different points in the course sequence, such as a summer precalculus option between 11th and 12th grade, opening up access to AP Calculus without having to compress or double up; this has its own complexities, including that 11th-grade applicants to 4-year universities do not yet have an advanced course reflected on their transcripts.

The big idea here is not to argue that you offer the same ways to accelerate as SFUSD has, or even that you layer in these options over time as SFUSD has. Rather, you should understand that you will need to make adjustments to whatever you initially design. The SFUSD team knows their students are currently accessing each of the four district-sponsored options and that it is a larger and more diverse group of students accessing these options. Figure 10.4 shows an increase across all groups of students earning 30 credits of math, or 3 full years of math credit, by the end of their junior year. Earning these credits not only shows that a student is on track to graduate but also is on track to take an advanced math course in their senior year. You can see that the trend line is up and that SFUSD still has disparities across groups of students.

FIGURE 10.4   STUDENTS MEETING MATH REQUIREMENTS

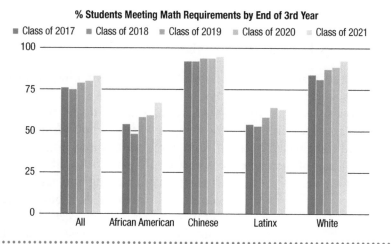

**% Students Meeting Math Requirements by End of 3rd Year**

■ Class of 2017   ■ Class of 2018   ■ Class of 2019   ■ Class of 2020   ▢ Class of 2021

Source: San Francisco Unified School District.

Another indicator of access is shown in Figure 10.5. This stacked bar graph shows the ways in which students are accessing more and different acceleration options, with about 40%, or 1,400 students per grade level cohort, accelerating in high school within one of the district-sponsored options. You can see, as one of these metrics, that the Accelerated Summer Geometry option was introduced first for the class of 2020. You can also see that students were steadily accessing more and varied paths to acceleration and that that progress slowed rather dramatically during the pandemic, with the class of 2023 not choosing options like taking two math courses during the period of distance learning.

FIGURE 10.5   STUDENTS ENROLLED IN ACCELERATION OPTION

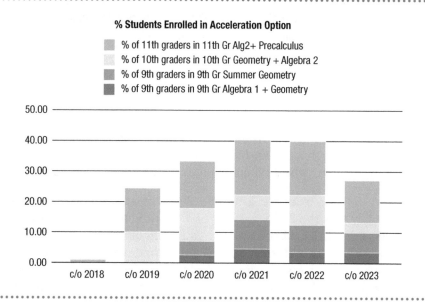

**% Students Enrolled in Acceleration Option**

▢ % of 11th graders in 11th Gr Alg2+ Precalculus
▢ % of 10th graders in 10th Gr Geometry + Algebra 2
■ % of 9th graders in 9th Gr Summer Geometry
■ % of 9th graders in 9th Gr Algebra 1 + Geometry

Source: San Francisco Unified School District.

Having more options and removing barriers also provide for a release valve when students and families are seeking many ways to access opportunities. At the time of this publication, we are currently looking at student outcomes in each pathway, ranging from grades in the courses themselves, to which courses students take next, and how they fare in those following courses.

Acceleration and tracking are not the same thing. Tracking sorts students into qualitatively different pathways, where some are and some are not on track to go to a 4-year institution—where the *smart kids* get treated as those to preference and accelerate and the *average kids* or *low kids* are left to lag, or to fend and fight based on their own community's and family's resilience and understanding of institutional education. We as authors believe that the literature of tracking and acceleration are conflated. Tracking does harm, and acceleration can provide access if the classes are kept as heterogeneous as possible. Educators, journalists, researchers, parents, and activist groups need to stop conflating these terms and treating them as synonymous.

> *Acceleration and tracking are not the same thing. Educators, journalists, researchers, parents, and activist groups need to stop conflating these terms and treating them as synonymous.*

## LESSON LEARNED

There is no simple solution to developing acceleration options, especially for students who wish to take AP Calculus in high school. Whatever you develop, hold to the core beliefs of your policy without being inflexible or dogmatic.

## When Policies External to Your Context Require You to Shift

Policies outside of your immediate sphere of influence may require you to shift what you are designing and how you are enacting that design. While you will need to be flexible and innovative, you should expect to have to change and not be surprised when changes come.

A California-specific example after the 2014 passage of SFUSD's Course Sequence Policy was the 2015 passage of a Math Placement Act, SB 359. As a unified PK–12 district, SFUSD had already articulated placement practices as students moved through their schools; they had also established a placement practice for students entering SFUSD high schools from private, charter, or other public schools where they may have taken high school Algebra 1 in middle school. SFUSD knew that many private institutions were not implementing the new standards, in many cases using textbooks that dated back to the old CA 1997 standards, often teaching content in procedural ways. Students who may have taken a course called "Algebra 1" would sit for a Math Validation Test (MVT), which would allow them to demonstrate mastery of Common Core Math 8 and Algebra 1 on a task-based assessment. Meeting the cut score on this exam would mean that a student newly enrolling in SFUSD in 9th grade would be enrolled in Geometry. Again, this predates the passage of SB 359. With the passage of the 2015 Math Placement Act, all districts were required to have a fair, objective, and transparent math placement process within the first month of a student's 9th-grade year. One equity-based aim of this bill was to interrupt some of the inequitable ways that students from different middle schools or feeder districts were being differently sorted as they matriculated into high school. When SB 359 was passed, SFUSD was able to make a 2016 addendum to their 2014 policy and leverage this existing exam, which was already designed as a check for mastery of the same relevant math content; these students would also need a course transcript of the UC approved Algebra 1 course they had taken.

The exam itself was designed with purpose and intent, alongside a rigorous scoring process; the tasks come from the Mathematics Assessment Collaborative (MAC), which had grown over decades as an assessment consortium of more than 100 districts.

> MAC is formed for the purpose of producing, scoring, and reporting student mathematical performance assessments at grades two through the first three years of college prep high school mathematics. . . . [The] method used to establish cut scores was 'developed jointly by the National Council on Measurement in Education (NCME), the American Education Research Association (AERA), and the American Psychological Association (APA). (Callahan Consulting, 2017)

Though not originally designed for SFUSD families who chose to pay for Algebra coursework—such as an online course outside of SFUSD—the MVT matched this new state requirement as well.

To be clear, this is a very small number of SFUSD 8th-grade families paying to enroll in an online Algebra 1 course outside of SFUSD. More than 1,500 students accessed SFUSD–sponsored acceleration options in their high schools and centralized summer programming in 2021, with these options free to families; by comparison, only 59 total administrations of the MVT were taken by SFUSD students that same year, meaning that less than 4% of families looking to accelerate were paying for coursework outside of SFUSD. Less than 4% of SFUSD families were paying for something private outside of their own schools, and it was still an important piece of legislation to respond to and take seriously.

The big idea here is not to argue that you design a validation test as SFUSD did, no matter how strong that assessment design; in fact, SFUSD has moved away from placement tests in general. Rather, the point is that outside policy shifts will likely necessitate that you make adjustments to whatever you initially design. This is the story of how SFUSD fielded this one narrow slice of its overall policy monitoring and adjusting.

As you think about any proposed policy changes, what is the core of that policy? Is the policy flexible enough to allow for iteration, without losing the core of your beliefs? If you want to explore these concepts, there are many places to begin:

- Look into the state and local policies about mathematics, including but not limited to placement practices.

- Check in on what is required of your school and district regarding math placement. Where is there space to move toward equitable pathways?

- Make community-friendly fact sheets, FAQs, and visuals that describe your proposed adjustments.

- If and when you are making further adjustments to your ongoing design, remember to update these resources and share the updates in whatever avenues you have.

- If you are in a school or district, do an inventory of the math-related administrative regulations in the middle school and high school counselors' manuals. These may need to be updated over time, as well as any professional guidance you can offer.

## LESSON LEARNED

Design for equitable outcomes. Be ready to make adjustments as your context changes, so long as you can still hold on to the core purpose of your original design for equitable math placement.

## FAMILIARIZING AND BUILDING CAPACITY OF NEW MEMBERS TO YOUR COMMUNITY

One of the lessons the SFUSD team learned, and named directly in Chapter 4, is to be mindful that changes in governing boards and senior leaders can manifest huge shifts in your policy landscape. We ask you to be ready to start at the beginning, as often as needed, and remember that all systems change will have forward movement and backward movement. The same kinds of statements are also true for teachers new to the profession or to your school or district, for site leaders new to their role or to your district who may never have taught math themselves, and for content specialists and coaches new to your site or your central team. It is natural for each of these educators to have foundational questions, especially given how sensitive conversations about math course sequencing continue to be. If you are going to sustain the implementation of bold equity-based work, you will need a plan to support training new staff and leaders and developing their capacity.

## Supporting People New to the Work

Many organizations are skilled with *onboarding* staff regarding the logistics: This is your new email, this is how we organize our shared folders, this is our open enrollment date for health benefits, and so on. Many organizations are far less skilled with bringing people into the big ideas of mission and vision, both in terms of sharing past practice, and engaging new staff who will also bring their own ideas, spheres of influence, and ways of working. Sometimes, this kind of onboarding may not be seen as having the same urgency as the logistics of getting someone a paycheck, so it moves to the back burner. It may not be obvious to an organization that new staff cannot immediately see and understand how your policies fit together to make a whole. All organizations will have some variability in where this is done well and where this is done less well.

In SFUSD, during the 2017 transition from then interim superintendent to the newly appointed superintendent Dr. Vincent Matthews, one of the first conversations the head academic officer knew he needed to talk through with Dr. Matthews was SFUSD's overall math program—specifically the secondary course math policy. In that first meeting, the head academic officer shared a series of things: the slides from the most recent formal presentation the math leadership team had shared with the Board of Education, the questions they had been asked and which they expected people would also ask of him, the questions the team still had, and what they were trying to get better at. While Dr. Carranza was on the ground as the team first passed the policy and during the first wave of implementation, Dr. Matthews was learning about multiple initiatives underway in SFUSD, as well as getting to know SFUSD's culture, structure, and individual people and teams. He wanted to understand how the policy was connected to other work in the district.

The same is true of teachers new to a site, who will need to become familiar with a school's practices and beliefs. At some of SFUSD's high schools that have long been a part of equity work, the math teachers advocate to be on interview panels for new teachers and to include direct questions about equity in mathematics and teaching in heterogeneous classrooms. Many of these candidates interview at the same site where they have studied teaching, paired through credential programs that know about and support SFUSD's work in heterogeneous classrooms. Early career teachers hired into these schools know "this is what it means to work here." Similarly, one of the ways SFUSD onboards content specialists is to ask that they themselves take the Complex Instruction (CI) course; they take it as a participant alongside schools new to CI and teachers new to CI schools. This supports members new to the team to have a shared experience in service of a shared vision of what is possible in heterogeneous classrooms.

## LESSONS LEARNED

As you bring people along who are new to your district or school, or new to their role, remember that calling people in is as much about culture and beliefs as it is about logistic aspects of being an employee. Each educator newly learning about your detracked course sequence and heterogenous math instruction will need help to understand the why, the what, and the how.

## Integrating Ideas

An additional layer often not named alongside the logistic onboarding of individual people is how to integrate initiatives so that they are complementary and experienced as coherent. As an example, like many school districts, in 2013, SFUSD developed a graduate profile that named the competencies expected of any graduate from a district (Kay, 2017). Because the Math Course Sequence Policy predated fuller descriptions the district developed for these graduate profile competencies, it was crucial for the math leadership team to be able to point to the profile as teachers and leaders themselves were making sense of it and say, "I see our work to support collaboration in heterogeneous classrooms reflected in this graduate profile attribute." For example, here is language from SFUSD's (2022) graduate profile that describes the attributes of leadership and empathy: "We organize activities in teams to help students learn to work together. We encourage students to work with partners outside the classroom, too—such as family members or mentors—to prepare them to lead and collaborate." As a district, there was alignment between the graduate profile and what was already in place. It made sense how these things belonged together. This kind of intentional alignment will also help you ward off the initiative fatigue that so many educators face.

Similarly, a new initiative might require you to shift your approach to your policy and its implementation. One such example for SFUSD has been the district's shift to Deeper Learning, including but not limited to assessment for learning (Berger et al., 2014; Watkins et al., 2018). As described in

Chapter 5 about curriculum that supports heterogeneous instruction, the math leadership team had already thought about formative assessment as integral to its curriculum, where teachers are looking throughout a math unit at student thinking and adjusting instruction. However, assessment for learning also asks that students set and monitor their own goals. SFUSD does have some schools engaged in very deep practice with students monitoring their own learning, such as math notebooking, within a set of schools using lesson study as the learning model for teacher collaboration (McAnelly, 2021). However, the district as a whole recognizes that it can move to more fully embrace the idea that the students themselves can be co-designers who share their own goals and thinking in its more than 100 schools. This shift toward assessment for learning has already begun to flow into more intentional work in professional learning spaces. A detracked system of math teaching and learning can get better because of alignment with these kinds of systemwide shifts.

## LESSONS LEARNED

In a climate of initiative fatigue, look for ways to connect the salient points across multiple initiatives. Help educators, families, and decision makers in your school or district see the connective tissue.

How will you support educators new to your policy? How will you shift to integrate initiatives that will be introduced before, during, and after your policy shifts and ongoing work toward heterogeneous math instruction? If you want to explore these concepts, there are many places to begin:

- Think about how you already onboard staff. If your onboarding focuses more on logistics than organizational beliefs and values, consider how you can shift the balance to be more about who you are and less about how you behave.

- Shift interviewing practices to make it clear that you are moving toward equitable outcomes in math with detracked math sequences and heterogeneous math classes.

- As you move toward equity-based policies and practices, help potential candidates understand *what it means to work here.*

- Inventory the initiatives that are already in place in your system, including those that are not math-specific.

- Pay attention as new initiatives come online and get integrated into the identity of your school or district. Look for ways to connect these initiatives to detracked math sequences and heterogeneous math instruction so that they don't feel like separate, stand-alone volumes in a library of documents.

## USING LARGER MOVEMENTS TO SUPPORT YOUR LOCAL WORK

As you read this book, you benefit from a national conversation about detracking that you can and should leverage. The math leadership team in San Francisco is experiencing this as an exciting groundswell, because it was not the case for them when they were first developing policy recommendations, though they did have decades of research and best practice they could lean on. For example, you can look to nationally recognized documents, position papers, and policy briefs such as

- The *Catalyzing Change: Initiating Critical Conversations* series (NCTM, 2018). These resources together offer guidance for PK–12 to interrupt pervasive systems of tracking. If you have money for just one book, buy this one.

- Joint statements about teaching Calculus from the Mathematical Association of America (MAA) and the National Council of Teachers of Mathematics (MAA & NCTM, 2012)

- *Closing the Opportunity Gap: A Call for Detracking Mathematics,* a position paper from NCSM, Leadership in Mathematics (2020)

You also benefit from greater clarity from postsecondary institutions. For example, in 2016, the University of California Board of Admissions and Relations with Schools (BOARS) released a statement that included this language:

> BOARS *also strongly urges students not to race to calculus at the cost of full mastery of the earlier math curriculum. BOARS commends the Common Core's goal of deeper understanding of the mathematical concepts taught at each K–12 grade level.*

Other postsecondary institutions are also clarifying admissions language to help with the articulation from the K–12 schools into colleges and universities. Harvard's (2022) admissions language, while not a change to their placement practices, offers a compelling counter to the pervasive narrative that a student is not college-ready without having taken AP Calculus in high school.

> *Specifically, calculus is neither a requirement nor a preference for admission to Harvard. We understand that many students have no intention to pursue college coursework that requires a knowledge*

*of calculus, and that other students are unsure of their future college studies. We also understand that not all students have the same opportunities to take certain math classes in high school, including calculus. Thus, we encourage applicants to pursue the pathways through math that are available to them and aligned with their interests and goals.*

To be clear, neither of these prestigious universities is saying that no students should accelerate to take AP Calculus while in high school. Indeed, both make the case that students whose goals include STEM fields can and should pursue coursework that aligns with their goals, including AP Calculus if available. Rather, these statements and others are clarifying that Calculus is not an admissions requirement; these statements are supporting you to have real conversations with communities that still believe that pushing content as early as possible will make their child stand out in the application process.

More work is being done at the intersegmental level, pulling together educators from the K–12 schools, the community colleges, the state university systems, and partners; examples include the recent Conference Board for Mathematical Sciences series of forums titled "High School to College Mathematics Pathways: Continuing to Support State Effort for Mathematics Alignment," which pulled together delegations from across the nation to think about ways to clarify and better support alignment between each state's K–12 schools and postsecondary institutions. Legislation can also support you in this conversation. One example from California is Assembly Bill 705, which took effect in 2018. AB 705 is designed to support California community colleges to place students in transfer-level courses in both English and math so that students do not get stuck in a remedial math track and never earn the qualifications to transfer to a 4-year university.

In essence, as you progress along your journey, understand that you are not alone. There are many resources you can use to situate your school, your district, or your state in a national conversation. Being able to point to a credible institution outside of yourself will support you in having hard conversations.

What is the greater landscape and the larger national conversation? If you want to explore these concepts, there are many places to begin:

- Find and read one of the resources named throughout this book, including the various position papers named in this section of this chapter.

- Consider who might be most impacted by learning about these ideas with you.

- Find out about the admissions practices in the community colleges and universities that are the closest and most relevant to your students.

- Find out about legislation that could impact your work toward equitable math pathways.

- Consider possible next steps, such as applying for funding associated with legislation, or advocating to the decision makers associated with that legislation.

- If you are wondering how to find out about things like admissions practices or legislation, your local chapter of the NCTM or other math advocacy groups can likely share about the advocacy work they are doing local to your context.

## LESSON LEARNED

There is a groundswell of interest in detracking work across the United States and Canada. While it may take different forms, and there will always be steps forward and steps back, we are all together in this work.

## Questions to Consider as You Consider Your Ongoing Policy Landscape

▶ This final chapter explored a few essential questions, organized around the big ideas related to monitoring and maintaining your detracked math policy. Here, we want to bring these same questions back to you as you consider your own context.

- Where are you in the development and implementation of your policy? What will you consider as early indicators of your policy's effectiveness?

- What adjustments could you make and still hold to your core beliefs and values?

- How will you adjust as inevitable change impacts your work?

- How can you leverage the national conversation about detracking?

## CLOSING THOUGHTS

We want to leave you with a few closing thoughts. Wherever you are in your journey, there are many people who came before you and many who will come after. You will need to build a critical mass of colleagues who are ready to join this transformative and necessary work so that you experience this difficult journey as a collaboration. Together, you will experience some resistance and some pushback. You may encounter news reports about your efforts that are skewed, biased, or lack the depth needed to convey your goals and your plan. You may speak with individuals or groups whose narrow concept of mathematics education limits their appreciation for the expansiveness of and the shifts needed to support heterogenous classrooms. You will likely hear people appropriate the language of equity as they call for tracked systems of course placement. You will certainly face well-intentioned community members who do not see your goals as serving the specific children they care most about and who will actively oppose your efforts. Together, you will need to recommit again and again to find your own resilience and stay the course and continue to clarify, educate, and advocate so that all students in your care ultimately have advantages and access to the highest levels of math that match their goals—access and advantages traditionally reserved only for some. Systems are by nature resistant to change, and this work won't happen easily or automatically. But we cannot simply wait for someone to make detracking a national decree. We must do what we know advances the mission of educational equity and start the work internally and intentionally. In June Jordan's often quoted call to action, "We are the ones we have been waiting for."

## Activity 10: A Culminating Activity for This Book

▶ In Chapters 1 through 9 of this book, all the activities we offered to close each chapter were examples of things the math leadership team did in San Francisco, from ways they brought math to different spaces to ways they helped leaders at different levels make sense of the current state and their spheres of influence. With this final chapter, we want to encourage you to go back to one of the big ideas, to one of the list of things to try, or to an activity. Choose one that resonates either because of where you sit in your organization, or because it feels like something brave you have yet to try. Find a colleague or a group of colleagues, and workshop whatever it is you have chosen and why.

_____

_____

# REFERENCES

Achieve the Core. (2013). *Progressions documents for the Common Core State Standards for Mathematics.* https://achievethecore.org/page/254/progressions-documents-for-the-common-core-state-standards-for-mathematics.

Adiredja, A. P., & Louie, N. (2020). Untangling the web of deficit discourse in mathematics education. *For the Learning of Mathematics, 40*(2), 42–46.

Alloway, M., & Jilk, L. M. (2010). *Supporting students by supporting teachers: Coaching moves that impact learning.* Paper presented at the 32nd annual meeting of the North American Chapter of the International Group for the Psychology of Mathematics Education, Columbus, OH.

Attebury, A., Lacour, S. E., Burris, C., Welner, K. G., & Murphy, J. (2019). Opening the gates: Detracking and the international baccalaureate. *Teachers College Record, 121*(9), 1–63.

Baker, B. D., Oluwole, J., & Green, P. C., III (2013, January 28). The legal consequences of mandating high stakes decisions based on low quality information: Teacher evaluation in the race-to-the-top era. *Education Policy Analysis Archives, 21*(5). https://doi.org/10.14507/epaa.v21n5.2013

Baldinger, E. M. (2014). *Learning together: Looking for learning in coach-teacher interactions.* Paper presented at the Research Conference of the National Council of Teachers of Mathematics, New Orleans, LA.

Ball, D. L. (2018). The American Educational Research Association Presidential Address; just dreams and imperatives: The power of teaching in the struggle for public education [Video]. YouTube. https://www.youtube.com/watch?v=JGzQ7O_SIYY

Benjamin Banneker Association, The. (2017). Implementing a social justice curriculum in mathematics. http://bbamath.org/wp-content/uploads/2017/11/BBA-Social-Justice-Position-Paper_Final.pdf

Berger, P., Rugen L., & Woodfin, L. (2014). *Leaders of their own learning: Transforming schools through student-engaged assessment.* John Wiley & Sons.

Berry, R. Q., III, Conway, B., Lawler, B., & Staley J. (2020). *High school mathematics lessons to explore, understand, and respond to social injustice.* Corwin.

Boaler, J., & Staples, M. (2008). Creating mathematical futures through an equitable teaching approach: The case of Railside School. *Teachers College Record, 110*(3), 608–645.

Bryk, A. S. (2010). Organizing schools for improvement. *Phi Delta Kappan, 91*(7), 23–30.

Bryk, A. S., Gomez, L. M., Grunow, A., & LeMahieu, P. G. (2015). *Learning to improve: How America's schools can get better at getting better.* Harvard Education Press.

Bryk, A. S., Sebring, P. B., Allensworth, E. L., Luppescu, S. S., & Easton, J. (2010). *Organizing schools for improvement: Lessons from Chicago.* University of Chicago Consortium on Chicago School Research.

Burris, C. C., Heubert, J. P., & Levine, H. M. (2006). Accelerating mathematics achievement using heterogeneous grouping. *American Educational Research Journal, 43*(1), 137–154.

Burris, C. C., Wiley, E., Welner, K. G., & Murphy, J. (2008). Accountability, rigor, and detracking: Achievement effects of embracing a challenging curriculum as a universal good for all students. *Teachers College Record, 110*(3), 571–607.

Cabana, C., Shreve, B., & Woodbury, E. (2014). Building and sustaining professional community for teacher learning. In N. S. Nasir, C. Cabana, B. Shreve, E. Woodbury, & N. Louie (Eds.), *Mathematics for equity: A framework for successful practice* (pp. 175–184). Teachers College Press and National Council of Teachers of Mathematics.

California Mathematics Council. (2021). CMC's commitment to anti-racism. https://www.cmc-math.org/anti-racism

Callahan Consulting. (2017, January 25). *Validity and reliability report.* San Francisco Unified School District. https://www.sfusd.edu/mvt

Campbell, C. (2021). Educational equity in Canada: The case of Ontario's strategies and actions to advance excellence and equity for students. *School Leadership & Management, 41*(4–5), 409–428.

Carter, P. L., & Welner, K. G. (Eds.). (2013). *Closing the opportunity gap: What America must do to give every child an even chance.* Oxford University Press.

CAST. (2018). The UDL Guidelines (version 2.2). https://udlguidelines.cast.org/

Childress, S., Elmore, R. F., Grossman, A., & Moore Johnson, S. (2007). *Managing school districts for high performance.* Harvard Education Press.

City, E. A., Elmore, R. F., Fiarman, S. E., & Teitel, L. (2009). *Instructional rounds in education: A network approach to improving teaching and learning.* Harvard Education Press.

Clandfield, D., Curtis, B., Galabuzi, G.-E., San Vincente, A. G., Lingstone, D. W., & Smaller, H. (2014). *Restacking the deck: Streaming by class, race, gender in Ontario schools.* Canadian Centre for Policy Alternatives.

Coburn, C. E., & Stein, M. K. (Eds.). (2010). *Research and practice in education: Building alliances, bridging the divide.* Rowman & Littlefield.

Cogan, L. S., Schmidt, W. H., & Wiley, D. E. (2001). Who takes what math and in which track? Using TIMSS to characterize U.S. students' eighth-grade mathematics learning. *Education Evaluation and Policy Analysis, 23*(4), 323–341.

Cohen, E. G. (1994). *Designing groupwork: Strategies for the heterogeneous classroom* (2nd ed.). Teachers College Press.

Cohen, E. G., & Lotan, R. (1994). *Designing groupwork: Strategies for the heterogeneous classroom* (2nd ed.). Teachers College Press.

Cohen, E. G., & Lotan, R. A. (Eds.). (1997). *Working for equity in heterogeneous classrooms: Sociological theory in practice.* Teachers College Press.

Cohen, E. G., & Lotan, R. A. (2014). *Designing groupwork: Strategies for the heterogeneous classroom* (3rd ed.). Teachers College Press.

Common Core State Standard Initiative. (2021a). Key shifts in mathematics. Retrieved October 15, 2021 from http://www.corestandards.org/other-resources/key-shifts-in-mathematics/

Common Core State Standard Initiative. (2021b). Standards for Mathematical Practice. Retrieved October 15, 2021 from https://www.cde.ca.gov/be/st/ss/mathpractices.asp

Cowen, J. M., & Strunk, K. O. (2015). The impact of teachers' unions on educational outcomes: What we know and what we need to learn. *Economics of Education Review, 48*, 208–223.

Craig, J. S. (1995). Quality through site-based scheduling. *Middle School Journal, 27*(2), 17–22.

Cuban, L. (September 30, 2018). "What happened to detracking?" Larry Cuban on school reform and classroom practice. https://larrycuban.wordpress.com/2018/09/30/whatever-happened-to-detracking/

Curtis, B., Smaller, H., & Livingstone, D. W. (1992). *Stacking the deck: The streaming of working-class kids in Ontario schools.* Our Schools Our Selves.

Domina, T., Hanselman, P., Hwang, N., & McEachin, A. (2016). Detracking and tracking up: Mathematics course placements in California middle schools, 2003–2013. *American Education Research Journal, 53*(4), 1229–1266.

Domina, T., McEachin, A., Hanselman, P., Agarwal, P., Hwang, N., & Lewis, R. W. (2019). Beyond tracking and detracking: The dimensions of organizational differentiation in schools. *Sociology of Education, 93*(3), 293–322.

Dweck, C. S., & Yeager, D. S. (2019). Mindsets: A view from two eras. *Perspectives on Psychological Science, 14*(3), 481–496.

Elmore, R. F. (1995). Structural reform and educational practice. *Educational Researcher*, *24*(9), 23–26.

Estrada, P. (2014). English learner curricular streams in four middle schools: Triage in the trenches. *Urban Review*, *46*, 535–573. https://doi.org/10.1007/s11256-014-0276-7

Farrell, C. C., Penuel, W. R., Coburn, C., Daniel, J., & Steup, L. (2021). *Research-practice partnerships in education: The state of the field*. William T. Grant Foundation.

Featherstone, H., Crespo, S., Jilk, L., Oslund, J., Parks, A., & Wood, M. (2011). *Smarter together! Collaboration and equity in the elementary math classroom*. National Council of Teachers of Mathematics.

Finkelstein, N., Fong, A., Tiffany-Morales, J., Shields, P., & Huang, M. (2012). *College bound in middle school and high school: How math course sequences matter*. The Center for the Future of Teaching and Learning at WestEd.

Foster, D. (2019). *Why use math performance tasks, with David Foster and SVMI*. Educational Data Systems. https://eddata.com/2019/01/why-use-math-performance-tasks-with-david-foster-and-svmi/

Gamoran, A. (1997). The stratification of high school learning opportunities. *Sociology of Education*, *60*(3), 135–155.

Gamoran, A. (2021). In high school math, more instructional time helps, but the tracking dilemma remains. *Proceedings of the National Academy of Sciences*, *118*(29), e2109648118.

Gamoran, A., & Mare, R. D. (1989). Secondary school tracking and educational inequality: Compensation, reinforcement, or neutrality? *American Journal of Sociology*, *94*(5): 1146–1183.

Gamoran, A., Nystrand, M., Berends, M., & LePore, P. C. (1995). An organizational analysis of the effects of ability grouping. *American Educational Research Journal*, *32*(4), 687–715.

Goffney, I., Gutiérrez, R., & Boston, M. (Eds.). (2018). *Rehumanizing mathematics for black, indigenous, and Latinx students*. The National Council of Teachers of Mathematics.

Gutiérrez, R. (2008). A gap-gazing fetish in mathematics education? Problematizing research on the achievement gap. *Journal for Research in Mathematics Education*, *39*(4), 357–364.

Hallinan, M. T. (2006). The detracking movement: Why children are still grouped by ability. *Education Next*, *4* (4). https://www.educationnext.org/the-detracking-movement/

Hanushek, E. A. (1999). Some findings from an independent investigation of the Tennessee STAR experiment and from other investigations of class size effects. *Educational Evaluation and Policy Analysis*, *21*(2), 143–163.

Harvard College. (2022). Harvard application requirements. https://college.harvard.edu/admissions/apply/application-requirements

Henderson, A. T., Mapp, K. L., Johnson, V. R., & Davies, D. (2007). *Beyond the bake sale: The essential guide to family/school partnerships*. The New Press.

Hitchcock, C., Meyer, A., Rose, D., & Jackson, R. (2002). Providing new access to the general curriculum: Universal design for learning. *Teaching Exceptional Children*, *35*(2), 8–17.

Horn, I. S. (2006). Lessons learned from detracked mathematics departments. *Theory Into Practice*, *45*(1), 72–81.

Huguet, A., Coburn, C. E., Farrell, C. C., Kim, D. H., & Allen. A.-R. (2021). Constraints, values, and information: How leaders in one district justify their positions during instructional decision making. *American Education Research Journal*, *58*(4), 710–747.

Hutt, E., & Polikoff, M. S. (2020). Toward a framework for public accountability in education reform. *Educational Researcher*, *49*(7), 503–511.

Illustrative Mathematics. (2016). *Cup of rice*. https://tasks.illustrativemathematics.org/content-standards/6/NS/A/1/tasks/463

Jilk, L. M. (2016). Supporting teacher noticing of students' mathematical strengths. *Mathematics Teacher Educator*, *4*(2), 188–199.

Jilk, L. M., & O'Connell, K. (2014). Reculturing high school mathematics departments for educational equity and excellence. In N. S. Nasir, C. Cabana, B. Shreve, E. Woodbury, & N. Louie (Eds.), *Mathematics for equity: A framework for successful practice*. Teachers College Press.

Kalodrides, D., Loeb, S., & Beteille, T. (2012). Systemic sorting: Teacher characteristics and class assignments. *Sociology of Education*, *86*(2), 103–123.

Kay, K. (2017). *The graduate profile: A focus on outcomes.* https://www.edutopia.org/blog/graduate-profile-focus-outcomes-ken-kay

Kendi, X. I. (2019). *How to be an antiracist.* One World.

Koski, W. S., & Levin, H. M. (2000). Twenty-five years after *Rodriguez:* What have we learned? *Teachers College Record*, *102*(3), 480–513.

Koski, W. S., & Reich, R. (2007). When "adequate" isn't: The retreat from equity in educational law and policy and why it matters. *Emory Law Journal*, *56*(3), 545–617.

Lee, V. E., & Bryk, A. S. (1988). Curriculum tracking as mediating the social distribution of high school achievement. *Sociology of Education*, *61*(2), 78–94.

Lee, V. E., Smith, J. B., & Coninger, R. G. (1997). How high school organization influences the equitable distribution of learning in mathematics and science. *Sociology of Education*, *70*(2), 128–150.

Long, M. C., Conger, D., & Iatarola, P. (2012). Effects of high school coursetaking on secondary and postsecondary success. *American Education Research Journal*, *49*(2), 285–322.

Lopez, G., & Gartzman, M. (2016). *Investigating course pathways in mathematics. Executive summary: Findings from district and school staff interviews.* UChicago STEM Education at the University of Chicago.

Lotan, R. (2006). Teaching teachers to build equitable classrooms. *Theory Into Practice*, *45*(1), 32–39.

Loveless, T. (2013, March). *The 2013 Brohern Center report on American education: How well are American students learning?* Brookings Institution.

Mathematical Association of America (MAA) and the National Council of Teachers of Mathematics (NCTM). (2012). *A joint position statement of the Mathematical Association of America and the National Council of Teachers of Mathematics on teaching calculus.*

Mathematics Assessment Project. (2022). *Home.* https://www.map.mathshell.org/

McAnelly, N. (2021). *How student math journals help students process their learning.* https://www.edutopia.org/article/how-math-journals-help-students-process-their-learning

McEachin, A., Domina, T., & Penner, A. M. (2019). *One course, many outcomes: A multi-site regression discontinuity analysis of early algebra across California middle schools.* (EdWorkingPaper: 19–153). Annenberg Institute at Brown University. http://www.edworkingpapers.com/ai19-153

Mosteller, F. (1995). The Tennessee study of class size in the early school grades. *The Future of Children*, *5*(2), 113–127.

National Council of Teachers of Mathematics. (2000). *Executive summary: Principles and standards for schools mathematics.* https://www.nctm.org/uploadedFiles/Standards_and_Positions/PSSM_ExecutiveSummary.pdf

National Council of Teachers of Mathematics. (2014). *Principles to actions ensuring mathematical success for all.*

National Council of Teachers of Mathematics. (2015). *The impact of mathematics coaching on teachers and students.* http://www.nctm.org/Research-and-Advocacy/research-brief-and-clips/Impact-of-Mathematics-Coaching-on-Teachers-and-Students/

National Council of Teachers of Mathematics. (2018). *Catalyzing change in high school mathematics: Initiating critical conversations.*

National Council of Teachers of Mathematics. (2020a). *Catalyzing change in early childhood and elementary mathematics: Initiating critical conversations.*

National Council of Teachers of Mathematics. (2020b). *Catalyzing change in middle school mathematics: Initiating critical conversations.*

National Council of Teachers of Mathematics and TODOS (2016). Mathematics education through the lens of social justice: Acknowledgment, actions, and accountability. https://www.todos-math.org/assets/docs2016/2016Enews/3.pospaper16_wtodos_8pp.pdf

NCSM. (2020, Spring). *Closing the opportunity gap: A call for detracking mathematics.* https://www.mathedleadership.org/docs/resources/positionpapers/NCSMPositionPaper19.pdf

Oakes, J. (2005). *Keeping track: How schools structure inequality, (2nd ed.).* Yale University Press.

Oakes, J., Ormseth, T., Bell, R., & Camp, P. (1990). *Multiplying inequities: The effects of race, social class, and tracking on opportunities to learn mathematics and science.* RAND Corporation.

Oakes, J., Wells, A. S., Jones, M., & Datnow, A. (1997). Detracking: The social construction of ability, cultural politics, and resistance to reform. *Teachers College Record, 98*(3), 482–510.

Okonofua, J. A., Walton, G. M., & Eberhardt, J. L. (2016). Bias or empathy in universal screening? The effect of teacher–student racial matching on teacher perceptions of student behavior. *Perspectives on Psychological Science, 11*(3), 381–398.

Oregon Department of Education. (2022). Oregon Mathways Initiative, high school math pathways—2 + 1 Model. https://www.oregon.gov/ode/educator-resources/standards/mathematics/Documents/2%20+%201%20Model%20Feb%202022.pdf

Palarady, G. J., Rumberger, R. W., & Butler, T. (2015). The effect of high school socioeconomic, racial, and linguistic segregation on academic performance and school behaviors. *Teachers College Record, 117*(12), 1–52.

Parekh, G., Killoran, I., & Crawford, C. (2011). The Toronto connection: Poverty, perceived ability, and access to education equity. *Canadian Journal of Education/Revue Canadienne De l'éducation, 34*(3), 249–279.

Paris, D. (2012). Culturally sustaining pedagogy: A needed change in stance, terminology, and practice. *Educational Researcher, 41*(3), 93–97.

Penner, A. M., Domina, T., Penner, E. K., & Conley, A. (2015). Curricular policy as a collective effects problem: A distributional approach. *Social Science Research, 52*, 627–641.

Penuel, W. R., & Gallagher, D. J. (2017). *Creating research practice partnerships in education.* Harvard Education Press.

Podolsky, A., Kini, T., & Darling-Hammond, L. (2019). Does teaching experience increase teacher effectiveness? A review of US

research. *Journal of Professional Capital and Community, 4*(4), 286–308.

Quinn, R. (2020). *Class action: Desegregation and diversity in San Francisco schools.* University of Minnesota Press.

Riegle-Crumb, C. (2006). The path through math: Course sequences and academic performance at the intersection of race-ethnicity and gender. *American Journal of Education, 113*(1), 101–122.

Riegle-Crumb, C., & Grodsky, E. (2010). Racial-ethnic differences at the intersection of math course-taking and achievement. *Sociology of Education, 83*(3), 248–270.

Rose, D. H., & Meyer, A. (2002). *Teaching every student in the digital age: Universal design for learning.* Association for Supervision and Curriculum Development.

San Francisco Unified School District. (2021). *Our core values.* https://www.sfusd.edu/about-sfusd/our-mission-and-vision/our-core-values

San Francisco Unified School District. (2022). *Middle grades redesign initiative.* https://www.sfusd.edu/learning/new-approaches-learning/middle-grades-redesign-initiative

San Francisco Unified School District Mathematics Department. (2021a). *Math talks.* Retrieved October 15, 2021 from https://www.sfusdmath.org/math-talks.html

San Francisco Unified School District Mathematics Department. (2021b). *Math teaching toolkit.* Retrieved October 15, 2021 from https://www.sfusdmath.org/toolkit.html

San Francisco Unified School District Mathematics Department. (2021c). *Participation quiz/group work feedback.* Retrieved October 15, 2021 from https://www.sfusdmath.org/participation-quiz–groupwork-feedback.html

San Francisco Unified School District Mathematics Department. (2021d). *The 3-read protocol.* Retrieved from https://www.sfusdmath.org/3-read-protocol.html

Seda, P., & Brown, K. (2021). *Choosing to see: A framework for equity in the math classroom.* Dave Burgess Consulting.

SERP Media. (2014). *Phil Daro—Against answer getting.* https://serpmedia.org/daro-talks/

SFUSD.edu (2022). *Grade 3 math.* https://www.sfusd.edu/learning/curriculum/

elementary-school/mathematics/grade-3-math

SFUSD Graduate Profile. (2022). *The graduate profile.* https://www.sfusd.edu/about-sfusd/our-mission-and-vision/vision-2025/graduate-profile

Skinner, A., Louie, N., & Baldinger, E. M. (2019). Learning to see students' mathematical strengths. *Teaching Children Mathematics, 25*(6), 338–344.

Smith, M. S., & Stein, M. K. (2018). *5 practices for orchestrating productive mathematics discussions.* The National Council of Teachers of Mathematics.

Steele, C. M. (1997). A threat in the air: How stereotypes shape intellectual identity and performance. *American Psychologist, 52*(6), 613–629.

Stringer, E. T., & Aragon, A. O. (2020). *Action research* (5th ed.). SAGE.

Teaching for Robust Understanding Framework. (n.d.). TRU intro. Retrieved October 15, 2021 from https://truframework.org/tru-intro/

Thompson, K. D. (2017). What blocks the gate? Exploring current and former English learners' math course-taking in secondary school. *American Educational Research Journal, 54*(4), 757–798.

TODOS. (2022). TODOS: Mathematics for All Excellence and Equity in Mathematics. https://www.todos-math.org/

TODOS: Mathematics for All. (2020). The mo(ve)ment to prioritize antiracist mathematics: Planning for this and every school year. https://www.todos-math.org/assets/The%20Movement%20to%20Prioritize%20Antiracist%20Mathematics%20Ed%20by%20TODOS%20June%202020.edited.pdf

Tomlinson, C. A. (2014). *Differentiated classroom: Responding to the needs of all learners* (2nd ed.). Association for Supervision and Curriculum Development.

Torres, A., & Woodbury, E. (2021). *Supporting teacher leaders to advocate for systems change.* Activity adapted from SFUSD math Complex Instruction program. California Math Council North. Asilomar Conference Presentation.

Tyson, W., Lee, R., Borman, K. M., & Hanson, M. A. (2007). Science, technology, engineering, and mathematics (STEM) pathways: High school science and math coursework and postsecondary degree attainment. *Journal of Education for Students Placed at Risk (JESPAR), 12*(3), 243–270.

Umansky, I. M. (2016). Leveled and exclusionary tracking: English learners' access to academic content in middle school. *American Educational Research Journal, 53*(6), 1792–1833.

University of California Board of Admissions and Relations with Schools (BOARS). (2016, April). Statement on the Impact of Calculus on UC Admissions UC Board of Admissions and Relations with Schools (BOARS). https://senate.universityofcalifornia.edu/_files/committees/boars/documents/BOARS_Statement-Impact-Calculus.pdf

Walton, G. M., & Cohen, G. L. (2007). A question of belonging: Race, social fit, and achievement. *Journal of Personality and Social Psychology, 92*(1), 82–96.

Wang, J., & Goldschmidt, P. (2003). Importance of middle school mathematics on high school students' mathematics achievement. *Journal of Educational Research, 97*(1), 3–19.

Watanabe, M. (2012). *"Heterogenius" classrooms: Detracking math and science. A look at groupwork in action.* Teachers College Press.

Watkins, J., Peterson, A., & Mehta, J. (2018). *The deeper learning dozen: Transforming school districts to support deeper learning for all: A hypothesis.* https://static1.squarespace.com/static/5bae5a3492441bf2930bacd1/t/5c044d39758d469484888c20/1543786237346/Deeper+Learning+Dozen+White+Paper+%28Public+2%29.pdf

Weathers, E. S. (2019). Bias or empathy in universal screening? The effect of teacher-student racial matching on teacher perceptions of student behavior. *Urban Education.* https://doi.org/10.1177/0042085919873691

Weiss, C., & Bucuvalas, M. (1980). *Social science research and decision-making.* Columbia University Press.

WestEd. (2022). Math in common: Executive summary. https://www.wested.org/wp-content/uploads/2019/06/MiC-Executive-Summary.pdf

White, D. Y., Gomez, C. N., Rushing, F., Hussain, N., Patel, K., & Pratt, J. (2018). Assembling the puzzle of mathematical strengths.

*Mathematics Teaching in the Middle School,*
*23*(4), 268–275.

youcubed. (2021). Data science unit 6 and new detracking workshop! [Email]. https://drive. google.com/file/d/1QoHX9iUliG23j55Kvlp WY5DTlMUjCo3Y/view?usp=sharing

Zavadsky, H. (2009). *Bringing school reform to scale: Five award-winning urban districts.* Harvard Education Press.

Zavadsky, H. (2012). *School turnarounds: The essential role of districts.* Harvard Education Press.

# INDEX

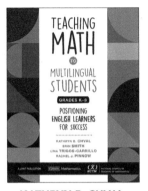
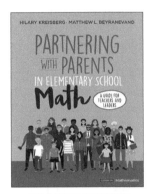
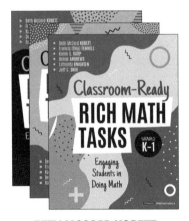
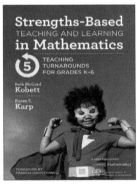

**JENNIFER M. BAY-WILLIAMS,
JOHN J. SANGIOVANNI, ROSALBA SERRANO,
SHERRI MARTINIE, JENNIFER SUH**

Because fluency is so much more
than basic facts and algorithms

Grades K–8

**KAREN S. KARP,
BARBARA J. DOUGHERTY,
SARAH B. BUSH**

A schoolwide solution for students'
mathematics success

Elementary, Middle School, High School

**JOHN J. SANGIOVANNI, SUSIE KATT,
LATRENDA D. KNIGHTEN, GEORGINA RIVERA,
FREDERICK L. DILLON, AYANNA D. PERRY,
ANDREA CHENG, JENNIFER OUTZS**

Actionable answers to your most
pressing questions about teaching
elementary and secondary math

Elementary, Secondary

**SARA DELANO MOORE,
KIMBERLY RIMBEY**

A journey toward making
manipulatives meaningful

Grades K–3, 4–8

CORWIN

A SAGE Publishing Company

Helping educators make the greatest impact

**CORWIN HAS ONE MISSION:** to enhance education through intentional professional learning.

We build long-term relationships with our authors, educators, clients, and associations who partner with us to develop and continuously improve the best evidence-based practices that establish and support lifelong learning.